Plough Quarterly

BREAKING GROUND FOR A RENEWED WORLD

Spring 2019, Number 20

Artists: Michael Naples, Sieger Köder, Carl Juste, André Chung, Ángel Bracho, Winslow Homer, Raymond Logan, Sybil Andrews, Cameron Davidson

Plough Quarterly

WWW.PLOUGH.COM

Meet the community behind *Plough*

Plough Quarterly is published by the Bruderhof, an international community of families and singles seeking to follow Jesus together. Members of the Bruderhof are committed to a way of radical discipleship in the spirit of the Sermon on the Mount. Inspired by the first church in Jerusalem (Acts 2 and 4), they renounce private property and share everything in common in a life of nonviolence, justice, and service to neighbors near and far. The community includes people from a wide range of backgrounds. There are twenty-three Bruderhof settlements in both rural and urban locations in the United States, England, Germany, Australia, and Paraguay, with around 2,900 people in all.

To learn more or arrange a visit, see the community's website at *bruderhof.com.*

Plough Quarterly features original stories, ideas, and culture to inspire everyday faith and action. Starting from the conviction that the teachings and example of Jesus can transform and renew our world, we aim to apply them to all aspects of life, seeking common ground with all people of goodwill regardless of creed. The goal of *Plough Quarterly* is to build a living network of readers, contributors, and practitioners so that, in the words of Hebrews, we may "spur one another on toward love and good deeds."

Plough Quarterly includes contributions that we believe are worthy of our readers' consideration, whether or not we fully agree with them. Views expressed by contributors are their own and do not necessarily reflect the editorial position of *Plough* or of the Bruderhof communities.

Editors: Peter Mommsen, Veery Huleatt, Sam Hine. Creative Director: Clare Stober. Designers: Rosalind Thomson, Miriam Burleson. Managing editor: Shana Goodwin. Contributing editors: Maureen Swinger, Susannah Black.
Founding Editor: Eberhard Arnold (1883–1935).
Plough Quarterly No. 20: The Welcome Table
Published by Plough Publishing House, ISBN 978-0-87486-287-4
Copyright © 2019 by Plough Publishing House. All rights reserved.

Scripture quotations (unless otherwise noted) are from the New Revised Standard Version Bible, copyright © 1989 the Division of Christian Education of the National Council of the Churches of Christ in the United States of America. Used by permission. All rights reserved.

Front cover: art by Michael Naples; image used with permission. Back cover: photograph reproduced by permission from Cameron Davidson. Inside front cover: "Das Mahl," MISEREOR-Hungertuch "Hoffnung den Ausgegrenzten" by Sieger Köder © MVG Medienproduktion, 1996.

Editorial Office	*Subscriber Services*	*United Kingdom*	*Australia*
PO Box 398	PO Box 345	Brightling Road	4188 Gwydir Highway
Walden, NY 12586	Congers, NY 10920-0345	Robertsbridge	Elsmore, NSW
T: 845.572.3455	T: 800.521.8011	TN32 5DR	2360 Australia
info@plough.com	*subscriptions@plough.com*	T: +44(0)1580.883.344	T: +61(0)2.6723.2213

Plough Quarterly (ISSN 2372-2584) is published quarterly by Plough Publishing House, PO Box 398, Walden, NY 12586.
Individual subscription $32 per year in the United States; Canada add $8, other countries add $16.
Periodicals postage paid at Walden, NY 12586 and at additional mailing offices.
POSTMASTER: Send address changes to *Plough Quarterly*, PO Box 345, Congers, NY 10920-0345.

From Farm to Feast

Dear Reader,

FOOD IS THE GREAT passion of our age. Whether intended to confer status, health, or pleasure, what's on the table holds the same exalted place in today's culture that art did in the Renaissance. Alice Waters is our Leonardo, Daniel Boulud our Raphael, Anthony Bourdain our Caravaggio. Food writing, a genre that barely mattered two decades ago, now does much to prop up the bottom line of media companies. Eight in ten American adults watch cooking shows.

This isn't necessarily bad. While food as art is easy to ridicule, it's a good thing when people rediscover the homey crafts of raising, preparing, and serving stuff to eat. The farm-to-table movement has brought renewed concern for the health of farms and of the natural world. Food writers such as Jonathan Gold of the *Los Angeles Times* have taught those with disposable money to cherish local diners and traditional immigrant cuisines – and perhaps even to get to know a truck driver or an immigrant in the process.

All the same, there's something fake about our culture's Instagram-fueled obsession with food. The intimacy it promises – with traditional ways of eating and living, with the crafts of butcher and baker, with the land and the soil – is counterfeit, a slickly marketed feeling of rootedness. The same consumer capitalism that has pushed the family farm to near-extinction and is engineering a mass exodus of the young from rural communities is quick to co-opt localist and fair-trade slogans. In coming decades, most farm families will likely be replaced by agricultural conglomerates with fleets of driverless tractors. Already now the toll is tragic: the high suicide rate among India's farmers rightly draws outrage, and even in the United States farmers take their own lives at twice the rate of military veterans.

This issue of *Plough* traces the connections between farm and food, between humus and human. According to the first book of the Bible, tending the earth was humankind's first task: "The Lord God planted a garden in Eden, in the east; and there he put the man whom he had formed" (Gen. 2:8). The desire to get one's hands dirty raising one's own food, then, doesn't just come from modern romanticism, but is built into human nature. For better or worse, food – how it's grown, how it's shared – makes us who we are.

"I'm gonna sit at the welcome table," proclaims a spiritual first sung by enslaved African-Americans. The song refers to the Bible's closing scene, the wedding feast of the Lamb described in the Book of Revelation, to which every race, tribe, and tongue are invited. To those who composed the song, the welcome table must have seemed a remote dream. But it was also a promise – a divine pledge of a day of freedom and freely shared plenty, of earth renewed and humanity restored.

In the case of food, the symbol is the substance. Every meal, if shared generously and with radical hospitality, is already now a taste of the feast to come.

Warm greetings,

Peter

Peter Mommsen, *Editor*

Readers Respond LETTERS TO THE EDITOR

On D. L. Mayfield's "What's the Good of a School?" Winter 2019: I agree with much that D. L. Mayfield has to say, with two important caveats: First, doing the best for others is not always synonymous with putting our children in a local public school. It need not be the case that every child in charter, private, or homeschool is subtracting resources from those in public school. Every local situation is going to be different; for example, concerned parents in Baltimore have been unable to hold back the initiative to arm school police despite intense advocacy. Not every "poorly rated" school is a lovable underdog in need of a few more concerned parents; some places really are so corrupt, unsafe, and deprived that it does them no good to bring in a more economically diverse set of students and their parents (and it will harm those students in the process).

Second, we do have a unique responsibility to our own children. Those with disabilities, those who are simply not learning in their current environments, or those who are bullied or otherwise traumatized at school might need to learn in a different way – and that's perfectly OK! I think that we can recognize this while still upholding the vision that Mayfield sets forth. If we prioritize "loving the poor" over and above the real needs of our own children, we will set ourselves – and them – up for unnecessary heartache.

Matthew Loftus, Litein, Kenya

As someone who, together with my wife, has successfully shepherded four children through the K-12 education system, I was very disappointed with the article "What's the Good of a School?" Early on we discovered that each of our children had different needs, and we continually made decisions – often very difficult ones for our family – for what was best for each child. Our children were homeschooled, and attended private, public, and religious schools. At certain times all four children attended different programs. We have never felt guilty for doing so. As an interracial family, we believe we provided a good model for parents – especially other minority parents – to advocate for their children, and not to allow schools to provide mediocre programs and poor quality instruction. When friends criticized us for "wanting a better program for our children because they were gifted," we responded: "no – we believe schools should provide these kinds of programs for all children."

Mayfield makes the mistake so many critics of public education make, which is to claim that the main role of the public schools is to rectify our society's inequalities. Not only is this impractical, but it can end up hurting the very students that need the most help; two recent examples are the Zero Tolerance programs and the federal NCLB Act. She also perpetuates the old stereotype about gifted students: that they can achieve in any educational setting. The new view is that these students have unique educational needs that must be met if they are to succeed.

I should be clear that providing what's best for each child does not necessarily mean selecting the highest-rated school. While we lived in the district of such a school, we chose for our children to attend a city high school (in Colorado, parents can choose a school outside their district if there is space) because we wanted them to have a diverse educational experience.   *Francis Wardle, Denver, CO*

(continued on page 7)

4   Plough Quarterly • *Spring 2019*

Book Launch:
The Heart's Necessities

What are the heart's necessities? It's a question Jane Tyson Clement (1917–2000) asked herself over and over, both in her poetry and in the way she lived. Her observation of the seasons of the soul and of the natural world have made her poems beloved to many readers, most recently singer-songwriter Becca Stevens, who has given Jane's poetry new life – and a new audience – as lyrics in her songs. This book interweaves Jane's best poems and the story of her life with commentary by Becca describing how specific poems speak to her own life, passions, and creative process.

Plough will be hosting several events to celebrate the launch of this book. On April 28, Becca will be joined by other artists and members of Jane's family at The Falcon in Marlboro, New York, for an evening of music and poetry readings. On May 1, she will perform and talk about Jane's life and poetry at the Rockwood Music Hall in New York City. *Plough* readers interested in attending these events can visit *plough.com/events* for more details.

Master Woodworker: Paul Sellers

If you are interested in woodwork and the beauty of craftsmanship with hand tools, you may already have heard about Paul Sellers or have seen his regular YouTube videos. Paul has been working with wood most of his life, and for many years has been on a campaign to bring hand tools back into the workplace. Originally from Stockport, England, he spent some years in Texas but has lived in Oxford for the last decade. He is an advocate of the idea that loving one's work is an essential part of human existence and considers himself an amateur because he would do what he does regardless of whether or not he was paid. "I've faced hard times in my life; who hasn't? Sometimes we look for recovery in an instant, sometimes just by keeping busy doing things, but then there are times when only creative handwork can fix the broken. I love that more and more people find peace in silent places of working." *paulsellers.com*

Paul Sellers in his workshop

Poets in This Issue: Luci Shaw and Richard Spilman

Luci Shaw was born in 1928 in London and has lived in Canada, Australia, and the United States. A graduate of Wheaton College, she became co-founder and later president of Harold Shaw Publishers, and since 1988 has been a Writer in Residence at Regent College in Vancouver. Author of eleven volumes of poetry, Shaw lives with her husband, John Hoyte, in Bellingham, Washington, where they are members of St. Paul's Episcopal Church. Her most recent collection is *Eye of the Beholder* (Paraclete, 2018).

Richard Spilman lives in Hurricane, West Virginia. He is the author of *In the Night Speaking* and of a chapbook, *Suspension*. His poetry has appeared in many magazines, including *Poetry, The Southern Review, Western Humanities Review,* and *New Letters.*

The Boy and the Bull

MAUREEN SWINGER

Photograph courtesy of the author

IT STARTED OUT as a character-building exercise. For two years running, our community's gentle Brown Swiss cow had had a female calf, expanding our milk herd

The author's son with his bull calf

to the grand total of three. But the November before last, we leaned over the barn gate to see a cream-colored, wobble-kneed bull calf. What to do with him?

At about the same time, we were pondering what to do about the head-butting young manchild in our own household. Could the kid take care of the calf?

The barn is just a couple hundred yards from our back door. Our son could help the little bovine toward a fulfilled life and productive end come slaughtering time next fall.

My husband pointed out that we're not really a pet family. Our son has never cared for anything that depended on him for sustenance on a regular basis. He lost interest in the rabbit after one week. (So did the rest of us. It found a more stable and loving home.)

Our son likes meat; he likes to sit out in the deer stand with his dad, and has helped turn a buck into venison roast and sausage. But now he's going to look a little creature in the eye, name it, nurture it, and know that it has a one-year expiration date? Will he be traumatized?

When we pitched the project, he was so excited that I had to cross out my concern. The sweet little bull, Sport, never had to wait long for attention. Our son was out of bed like a shot every morning, stamping into his boots to jog out and dispense grain, hay, and fresh water.

No, the calf didn't wait – but we did. The boy was never home in time for breakfast. Neighbors got used to hearing first the bellow of a growing calf calling for breakfast, then the bellow of a frustrated parent calling, "Breakfast time." A cheerful "Coming!" would float back on the breeze, followed five or ten minutes later by the cowboy himself, remembering, sometimes, to shed his mucky boots before digging into his own grain.

Maureen Swinger is an editor at Plough. *She lives at Fox Hill, a Bruderhof in Walden, New York.*

There was always a great explanation. Once, Sport managed to get his head stuck in the fence rails. Once a farmhand put him in a farther pasture, and our boy had to lug an overflowing bucket of grain through the near pasture, where the three big cows converged on him hungrily. Friendly, yes; gentle, yes; 1400 pounds each, yes. It's a little hard to argue in such circumstances. One day, the hayloft begged to be explored; another, the fence needed someone to balance on it.

None of these were excuses, just adventures, narrated with wide-eyed wonder over the cereal bowl. Our "Yes, but you need to get home in time for breakfast" was met with complete agreement, as it would be the following (late) morning.

By summer, Sport's light fawn coloring had deepened to rich brown, and his bellow sounded like a teenager trying out a cuss word while hoping his voice wouldn't break. He no longer cared for a friendly scratch between the horn buds. Still, farmer boy was loyal and proud of his charge, and chatted amicably with him while he chowed down his corn.

By September, Sport was a sturdy yearling. Dad told son not to climb in the pen anymore. October came around, and our son went out on the last morning to say goodbye. He'd gone by himself every morning, and he wanted to go by himself now. So I don't know what words passed between them. He came back sober, but not flustered.

A well-meaning neighbor asked him, "How will you feel after they butcher your cow?" He answered, after a slight pause, "Full."

I realized I needn't have worried.

The barbecue fed two hundred people, and folks were kind enough to thank him for his work. I love watching the quiet pride on his face when he remembers Sport. "Mom, he was a great little guy. And didn't that meat taste good?" ➤

Readers Respond *(continued from page 4)*

On John Thornton's "A Debt to Education," Winter 2019: Pastor Thornton's article reminded me of an experience I had several years ago when I sat on a small public university's scholarship committee. We did not have a lot of money to give away – I believe our largest award was only a few thousand dollars. Two young women were in the running for this award: the first lived at home, worked part-time, and commuted to school while the second lived on campus and was active in student groups that promoted social justice. At first, most of the committee wanted to give the award to the second student because she had devoted her time to volunteer work and also had accumulated more debt. I protested, noting that the first applicant could not be as involved with extracurricular activities because she worked and drove back and forth to campus, forgoing the residential college experience in order to save money. To my mind, the first student had made a great sacrifice up front and deserved to be rewarded. In the end, we wound up dividing the award. I do not think I was viewing the second student through a retributive lens; rather, I wanted the first young woman to know that her "bean counting," as Pastor Thornton puts it, and the subsequent sacrifices that go along with such austerity, deserved recognition.   *Rachel Rigolino, Highland, NY*

We welcome letters to the editor. Letters and web comments may be edited for length and clarity, and may be published in any medium. Letters should be sent with the writer's name and address to letters@plough.com. ➤

EDWIDGE DANTICAT

This Is My Body

Of
Food
and
Freedom

THE FATHER of a friend used to tell her, as she enjoyed what he considered a bit too much food, that she was slowly digging her grave with her teeth. It's not an original thought. Others have said it before him and will continue to say it after. As my friend and I used to reply to her father and others who echoed him, if not with the same words but the same sentiment, *we know.*

I often think of this supposed oral grave digging when I am with people for whom food is not an indulgence, people for whom there is no excess food, people for whom food seems dangerous. In the immigration detention centers that I have visited, for example, the subject of food often comes up. Many of the detainees see the terrible food they are fed at the most inconvenient hours – sometimes at four in the morning for breakfast and four in the afternoon for dinner – as yet one more way of punishing them.

The food would neither "stay up nor down," one woman told me fourteen years ago when I met her in the South Florida hotel that had been turned into a holding facility for women and children who'd come to Miami by boat from Haiti. These women either vomited this food, or it gave them diarrhea. Six of them lived in one hotel room. Some were forced to sleep on the floor, but the food felt like the most humiliating torment. Not only did these women have no control over what they were putting in their bodies, but it was also making them sick, and the sickness was further dehumanizing them.

During my teenage years, in the early 1980s, my parents used to take me along with my father's deacon group to visit Haitian refugees and asylum seekers at a detention center near the Brooklyn Navy Yard. The topic of food came up then as well. Back then, the men believed that hormones in the detention center food were making them grow breasts, a condition known as gynecomastia. "They're trying to make us into women, so we can be more docile," one man had told my father during one of our visits to the Brooklyn detention center.

In October 1987, thirty Haitian men who had been detained at Miami's Krome Detention Center filed a civil suit against the federal government claiming that they'd developed gynecomastia while at Krome. The only thing the men detained in Brooklyn and the men detained in Miami appeared to have in common, besides being Haitian and being in detention, was the institutional food.

The lawsuit revealed that the gynecomastia might have been caused by the detention centers' use of insecticides, particularly a type meant for animals, and Kwell, a harsh anti-scabies and lice cream, which was given to Haitian detainees to use daily as a body lotion. Other research, however, found clear links between diet and gynecomastia, and the men remained convinced that the detention center food had something to do with it. Despite all this, a jury found the government not liable.

MEALS EATEN in desperation or under distress of course end up being memorable. The choice of pre-execution meals generates so much interest that these meals are often mentioned, along with the final words spoken by the dying, in postmortem press conferences. The most legendary final meal

Edwidge Danticat is the author of many books, including the forthcoming collection of stories Everything Inside *(Knopf, August 2019). She is a 2009 MacArthur Fellow and a 2018 winner of the Neustadt International Prize for Literature.*

is the Last Supper, which is the great-great-great-grandfather of all final meals. We have no account of what else was consumed at the Last Supper besides unleavened bread and wine, which Jesus offered to his disciples – including the ones who would renounce and betray him, by saying, "Take, eat; this is my body." Then, "Drink, this is my blood."

Many of the men my father and his church friends visited in the Brooklyn detention center those many years ago were religious. As were some of the women I visited years later and have continued to visit whenever I am allowed into immigration detention centers. Many unaccompanied detained children carry with them a cross or a Saint Christopher medal, among other precious amulets, which are meant to protect them on their grueling journeys. Saint Christopher is said to have carried a small, vulnerable child across a raging river, a child who could have been any child in need, a stranger's child who turned out to be the Christ Child. Saint Christopher, too, was a migrant who ended up being imprisoned, and eventually executed.

Many of the children crossing deserts and raging rivers these days also start out with some food that has been carefully cooked to sustain them at least part of the way. Then, for the rest of the journey, both parents and children must trust that they will somehow encounter food, either to buy or to be gifted,

and water to drink to stay alive. This requires as much faith as hoping that there is a lamp beside a golden door, a door that is still open to "huddled masses yearning to breathe free."

ONE OF THE WAYS my immigrant parents tried to enmesh my brothers and me in American culture was to let us choose between eating pizza, fried chicken, or hot dogs on Friday nights after eating rice and beans, plantains, and other Haitian dishes every other night of the week. I never told them that I was already eating these "American" foods daily for lunch at school, because

Carl Juste,
Crushed,
Port-au-Prince,
Haiti

Image on previous spread:
André Chung,
Cane Cutter,
Havana, Cuba

Carl Juste,
A Day's Work,
Port-au-Prince,
Haiti

I feared that my parents, too, might feel that I was digging my grave with my teeth.

My mother liked to tell my brothers and me that *sak vid pa kanpe* (empty sacks don't stand) and *se sa k nan vant ou ki pa w* (only what's in your belly is yours). And to that we would also reply, *we know*. She often told us these things right before we went to someone else's house for lunch or dinner. The ultimate lesson in those maternal maxims and familial proverbs was to never show up somewhere too hungry. You never knew when they would

be ready to feed you, and you must not seem too famished, too desperate, too *empty* when they do. And if by any chance your arrival at someone's house happens to coincide with a meal to which you were not previously invited, you must refuse the food you are offered, even if you are starving. Otherwise it will seem as though you purposely showed up for that meal and that would make you seem calculatingly greedy, *visye*.

I think of all this, too, when I hear about people who have nothing in their stomach of

not keep living without doing something to protest the injustice of my treatment. They could lock me up for no reason and with no chance to argue my innocence. They could torture me, deprive me of sleep, put me in an isolation cell, control every single aspect of my life. But they couldn't make me swallow their food.

In July 2013, Yasiin Bey, the rapper and activist formerly known as Mos Def, agreed to be force-fed in a manner similar to the way prisoners on hunger strike were being force-fed at Guantanamo Bay. Bey was strapped to a feeding chair that resembled an electric chair. His hands and feet and head were placed in restraints. A nasal gastric tube was forced through his nose, down the back of his throat, and into his stomach, a process the US military called enteral feeding. As Bey wriggled and twisted in the chair – to whatever extent he could – tears ran down his face. He coughed. He grunted. He pleaded with his "jailers," who sometimes pressed the weight of their bodies on his chest and stomach, to stop.

"Please, please, don't," he begged.

After a minute or so, he was squirming and struggling so much that the tube fell out. The jailers put him in a choke hold to further restrain him and only stopped when he spoke as himself, as Yasiin Bey, and said, "This is me. Please stop. I can't do this anymore." Then he broke down and cried.

Had Bey been an actual prisoner, his jailers wouldn't have stopped until they were done force-feeding him. Those on hunger strike at Guantanamo Bay were fed like this twice a day, and for two hours each time. They then had a mask placed over their mouths while their bodies processed the liquid nutritional supplement, which was often Ensure. Back in their "dry" cells, which meant there was no water in those cells, they were observed closely

their choosing, people who have to completely rely on others to feed them, people who have no choice but to swallow food they despise, and people who are fed against their will.

Lakhdar Boumediene was a prisoner in Guantanamo from 2002 to 2009. In 2017, he wrote in the *New Republic* about being on hunger strike there:

> I am sometimes asked why I went on a hunger strike. *Did you want to die? Had you given up?* The answer is no. . . . I stopped eating not because I wanted to die, but because I could

to see if they were vomiting. If they vomited the supplement, they were returned to the restraining chair. Many prisoners urinated and defecated on themselves in the chair, as one might expect. The feeding tube, of course, makes it hard to breathe. Prisoners who were fasting during Ramadan, the Muslim holy month, were force-fed before dawn and after sunset. At the beginning of this year, detainees went on hunger strike in immigration detention centers in Texas, Miami, Phoenix, San Diego, and San Francisco, and were force-fed through nasal tubes on the orders of a federal judge. The force-feeding has led to constant vomiting and nosebleeds, according to family members.

SOMETIMES WHEN a person dies under mysterious circumstances, when a person passes away suddenly with no previous sign of illness, for example, the elders in my family will say that this person was "eaten." *Yo mange li.* They ate him. They ate her. The *yo* (they) who have done the eating is often a person or a group of persons of ill will who have deployed some destructive force to remotely kill someone else. We might never willingly offer ourselves to be *eaten* in this way, unless we are noble to the point of being sacrificial, or unless we feel we have no choice.

In the early 1990s, before it became a military prison where terrorism suspects are detained indefinitely, the Guantanamo Naval Base in Cuba was used to warehouse thirty-seven thousand Haitian asylum seekers who'd been intercepted by the US Coast Guard on the high seas after Haiti's first democratically elected president, Jean-Bertrand Aristide, was overthrown in a military coup. And because HIV-positive immigrants were banned from entering the United States at the time,

HIV-positive Haitian asylum seekers were held there for what must also have felt like an indefinite amount of time. Over two hundred HIV-positive Haitians, led by Yolande Jean, a mother of two and Haitian political activist, started a hunger strike on January 23, 1993, and continued the strike for ninety days. At that time Yolande Jean told American reporters: "We started the hunger strike so that this body could get spoiled and then the soul can go to God. Let me kill myself so my brothers and sisters can live."

In a letter addressed to her family, particularly to her sons Hill and Jeff, while she was on hunger strike in Guantanamo Bay in 1993, Yolande Jean wrote:

> *To my family,*
> *Don't count on me anymore, because I am lost in the struggle of life. . . . Hill and Jeff, you don't have a mother anymore. Realize that you do not have a bad mother, only that life took me away . . . Good-bye my children. Good-bye my family. We will meet in another world.*

Yolande Jean survived, but half of the HIV-positive hunger strikers died after they were finally released.

How many of us would have the courage to write a letter like that? And how many more mothers will have to write, or come close to writing, letters like this – how many must keep saying *this is my body, this is my blood, this is my son, this is my daughter, this is my hope, this is my dream, this was my life, this is what my life was supposed to be, this is what I believed was going to be my death* – how many of us must recite these pleas, these prayers, these laments, and these dirges, before we are brought to the table in communion, and are allowed to sit and eat in peace? ⮑

At the
Welcome Table

Breaking bread around the world:
Plough **asked eight friends in eight
places to share what hospitality
looks and tastes like.**

Brazil •	Claudio Oliver
New York •	Leah Libresco
Zimbabwe •	Elizabeth Mambo
South Korea •	Seonghee Kim
Michigan •	Cozine A. Welch Jr.
West Virginia •	Ellesa Clay High
Nicaragua •	Jairo Condega Morales
Germany •	Clemens Weber

Building Tables, Not Walls • Curitiba, Brazil

Claudio Oliver: We live, we are told, in a world of scarce resources but unlimited needs and wants. The world answers this problem with exclusion, walls, and violence. If we are Christians, however, we are called to believe that God has created an abundant world, to which he has set limits but which can provide abundant life for all. As a community, we recognize this abundant creation not by building higher walls, but longer tables for common meals and conversation.

Every Saturday we have two opportunities, lunch and supper, to offer our friends a big table, fantastic organic pasta, and products from our urban farm production. That has been our answer to the last two years of division in our country. A big spread, full of honest and delicious food, which celebrates the most ancient of Christian traditions: shared table, shared words.

Teachers, neighbors, and workers young and old sit together at our long table. We learn about one another and enjoy food, stories, and life, while we preserve the welcoming cheerfulness that Thomas Aquinas called *eutrapelia*. A simple plate of gnocchi can bring tears of gratitude when served with frankness; a cup of water and a smile can bring hope to a depressed visitor; new friends are made over a piece of genuine Brazilian carrot cake. Most importantly, those with broken lives can experience restoration when it comes embodied, really and concretely, in abundant lives and welcoming tables.

Claudio and Katia Oliver live and serve as pastors in an intentional community in Curitiba, Brazil; for twenty-five years Casa da Videira has been dedicated to following the steps of Jesus.

All photographs courtesy of the authors

Mulled Cider in Manhattan • New York, New York

Leah Libresco: Despite the suggestion of one of my guests, I did not whittle pegs and stake fish over an open fire for our party, even if Kristin might have. In January, friends (and friends of friends of friends) came over to celebrate the end of our five-month journey through the one thousand pages of *Kristin Lavransdatter*. Sigrid Undset's epic can feel intimidating, so we made a Facebook group and vowed to stick it out to the (spoiler!) plaguey end.

Our discussion as we read had been lively, but all online, so my husband and I invited anyone who could make it to our home on the Upper West Side of Manhattan when the read was over, for some in-person discussion. We were pleasantly surprised by how far people were willing to come. Folks crossed the river from New Jersey, took the train up from D.C., and one couple took a flight from Ohio. We welcomed them all with mulled cider – and a bottle of bourbon for added warmth.

Some of the people in our studio apartment were old friends, others were strangers, but all, as they introduced themselves, took a bite of the salty Norwegian licorice I'd bought for the party. "I can't believe this book club has hazing!" someone complained. Two attendees ran to the kitchen to spit it out, and one woman went back for seconds, then thirds. Happily, one of the guests – taking her inspiration from the first volume of *Kristin*, "The Wreath" – had brought a Bundt cake.

Leah Libresco Sargeant is the author of Arriving at Amen *and* Building the Benedict Option: A Guide to Gathering Two or Three Together in His Name.

Cooking *Sadza* • Negovano, Zimbabwe

Elizabeth Mambo: The name of my village is Negovano. In this celebration, we're enjoying a long-awaited reunion of family and friends. Some have traveled from America; others have lived here their whole lives. When we celebrate, everyone is welcome. We can always make the food go around. We start and end our gatherings with prayer and singing – lots of singing. And we don't sing without dancing!

At every such gathering, we always begin by cooking *sadza*, the staple food in Zimbabwe, using cornmeal ground from field maize. It cooks slowly over an open fire; there's an art to making it just the right consistency. *Sadza* is filling and goes well with the other dishes: rice with peanut butter sauce, butternut squash, collard greens and tomato salad, and chicken or goat meat in a gently spiced stew. We grow everything ourselves.

Times are still very hard in rural Zimbabwe; we worry about the future of our country. There have been several years of drought, and the government situation hasn't improved, even though the leadership has changed. But it's not in us to lose hope. God is always with us, and we have each other.

Elizabeth Mambo is a physical therapist and mother of three. She now lives in New York, but stays in touch with her family in Zimbabwe and visits whenever she can.

The Warm Table • Seoul, South Korea

Seonghee Kim: We sit around the table with my wife's brother's family, and their mother. It's shortly after we moved to Seoul from the house we had built in the countryside. The food is not particularly festive; the table is a zelkova board that I cut from a discarded tree trunk.

We share multigrain rice, kimchi, *namul muchim* (seasoned vegetables), bulgogi (seasoned beef barbeque), and *doenjang-jjigae* (fermented bean paste stew) – our typical daily menu. Knees and chopsticks bump into each other; our living room is small, and many people sit around the table. My wife holds the baby so her sister-in-law can eat.

It isn't just family members who sit around this table. We often have friends, classmates, and colleagues come to our house to share meals. In our parent's generation, serving a good meal to travelers was a part of life. It has become rare to see such gatherings. Incomes have increased dramatically in Korea, but the room in our hearts and schedules has decreased. I see our society traveling a rich but lonely way. But the memory of our parents who served meals to such a variety of people reminds us: we know the meaning of "the warm table." And it is that which we are trying to carry on. A meal is not just made of individuals. It becomes a living body that warms the participants, encouraging them and giving them strength.

Seonghee Kim works for Hansalim, a cooperative movement in South Korea that connects farmers and consumers to "save all living things."

First Meal • Madison Heights, Michigan

Cozine A. Welch Jr.: "The more colorful the meal, the better for you" is the adage. So I order my blueberry pancakes with bright red strawberries and sunshine bananas topped with whipped cream. My father does the same, sans cream, strawberries, and bananas. I'm clearly the health-conscious one.

It's little over an hour since my release from a twenty-year sentence for a crime I committed at seventeen. My father and I sit opposite each other on stiff maroon cushions in a booth at Bob Evans. We had been farther apart from each other in view and disposition when we'd eaten here in the nineties. The 240 months of suffering and growth since then have partly bridged the gulf.

The mood, as we sit here at my first meal, late morning sunshine glinting off our spotted silverware, is a strange mix of joy and sadness. Missing from this reunion is my mother. Sixty-eight days before my release, two days after the fifteen minute phone call in which I told her I'd be coming home, she died. Oddly, we don't feel incomplete. It's as if we both can see her there, in her usual spot, eating her colorful omelet, green bell peppers poking through canary yellow eggs.

Leaving the prison, my father's radio played: "I'll always love my mama. She's my favorite girl." It's still in my head as we sit here, thankful and mournful; laughing, then silent. We look to the empty space where her wheelchair would have been and smile through watery eyes.

Cozine A. Welch Jr. is a published poet, editor, and teacher at the University of Michigan in Ann Arbor.

3 Sisters and a Kilt Salad • Preston County, West Virginia

Ellesa Clay High: Winter has set in, yet cars slide over snow-crusted ruts toward my home, half a mile from paved road. This evening I welcome friends to celebrate some Appalachian hospitality and tenacity.

These West Virginia ridges persist, ancient, inhabited for thousands of seasons. Our blessing, partly in Shawnee, thanks God and connects us to the animals and plants that gave their lives so we can eat. On my stove bubbles a stew of the "three sisters" – corn, beans, and squash – an eastern woodland tradition.

Dean and Lois Christopher live on a farm that has sustained six generations. Lois has baked rolls the "old way" with potato water and sugar as the starter. "Nothing was wasted back then, not even the water used to boil potatoes," she recalls. These rolls twang with memory.

Mike Costello and Amy Dawson have driven about two hours from their farm with a "kilt greens" salad. Kale and mustard buds are massaged with hot bacon grease, "killing" them. Topped with squash seeds, salt-roasted radishes, Arkansas Black apples, and salt-rising bread croutons, the salad is dressed with a vinaigrette of herbs and hickory syrup.

At the table end sits my son, Clay Johnson, a decorated veteran, and his four-year-old daughter, Lilia Coreen. The future of this farm belongs to them.

After dinner and Susan Sauter's blueberry cornbread upside-down cake, we'll linger over coffee and more stories. I won't mind cleaning up, because friendship holds the meal together – friendship burnished as the old pewter plates that I wash.

Ellesa Clay High is a writer and Emerita Associate Professor in Native American literature and creative writing at West Virginia University.

Island Festival • Ometepe, Nicaragua

Jairo Condega Morales: This past December, families from the Catholic parish of San Jose del Sur, a village of nine hundred on the small volcanic island of Ometepe, celebrated Christmas differently. We gave a feast for seventy-two schoolchildren, many of whom regularly go hungry. There was a piñata, gifts, games, dancing, and generous helpings of pork – the Ometepe pigs, which roam freely around the island, provide a flavorful and nutritious meat.

"Ometepe, oasis of peace," runs a much-quoted song in praise of my native island in Lake Nicaragua. For centuries, Ometepe has lived up to its nickname. Even during Nicaragua's great bloodlettings – first the Sandinista revolution, then the Contra insurgency – there was no fighting on the island's soil.

That changed in the second half of 2018, as police and government-aligned paramilitaries have cracked down on protestors in the aftermath of popular protests. There have been shootings, beatings, and arrests. As a result of the repression, Ometepe's economy, which depended heavily on European ecotourists, has tanked, with hotels, restaurants, and mountain guide firms now shuttered. Many of the thirty-five thousand islanders – traditionally small farmers, like my own parents – find themselves without income for medicine and necessities. Among children, malnutrition is on the rise.

That's why our family and others wanted to express our love for the children of our village – they are the most vulnerable ones. At the end of a hard year, this feast was a moment of shared happiness – for the children, and for us.

Jairo Condega Morales is an Ometepe native who coordinates Plough's El Arado *program of free books and educational programs for schools in Nicaragua.*

Guests at Home • Berlin, Germany

Clemens Weber: The room is prepared, the tables set. There's meatloaf and mashed potatoes and sauerkraut. It's guest night at our little community house in Berlin. There are six adults, four children, and one long-term guest in our community. We're a small bunch. But the church of Christ can't just be there for itself. So we open our doors.

There's an excitement on guest night: Whom will God send today? We never know how many to expect. Sometimes it's two, sometimes fifteen. But there's always enough food.

While we eat, we talk. Often we read a short text. Dinner is followed by singing, sometimes a game, and always a short devotion. We take Jesus at his word: "Wherever two or three are gathered in my name, there I am in the midst of them." It's that simple: we can't do more than prepare the food and then open the doors in the expectation that God can speak to us through each guest who comes.

Tom, homeless, joins us regularly. "This is a wonderful place," he told us. Hans, who grew up without parents, says he is grateful to have found a family. Ricardo, a junkie on the mend, painted us a picture. The upper half is golden, the lower half is dark. Jesus – a silver path – joins the two halves.

It's a fitting expression of our evenings together: the whole thing is carried by the Spirit. We are grateful. And in our gratitude, we feel like guests in our own home. ⤳

Clemens Weber, with his wife and their children, has lived for the past ten years in a Christian grassroots community in Prenzlauer Berg, Berlin.

Digging Deeper

BOOKS TO STIR THE CULINARY SOUL

JANE SLOAN PETERS

Peter Mayle's ***French Lessons: Adventures with Knife, Fork, and Corkscrew*** (Vintage, 2002) is a culinary road trip through France. Mayle, an Englishman who describes his move to a two-hundred-year-old farmhouse in the Lubéron in *A Year in Provence*, now introduces readers to the small-town French love affair with food and drink. He writes with a gusto for the table, and a wit one might expect from a Brit living among the French. Chapters describe a frog leg festival in Vittel, the annual Foire aux Escargots in Martigny-les-Bains, and a wine tasting in Burgundy. There is also a chapter devoted to – obviously – a cheese festival, but another devoted to a *messe de truffles*, a Mass in honor of the truffle. Delightful characters abound, too: Régis, the supercilious lover of chicken; Sadler, another English food writer with a penchant for stinky cheese; and nameless table companions who argue about the best ways to cook and eat an omelet. *French Lessons* is the fruit of Mayle's "long hours with knife and fork and glass that I like to call research" – hours we may relish with him in spirit, if not in stomach.

Julia Child's ***My Life in France*** (Knopf, 2006; with Alex Prud'homme) chronicles the adventures of America's most beloved chef during her years in Paris, Marseille, and Provence from 1948–1954. Child, writing at 92 years old, is a high-spirited and indefatigable narrator, charming readers with stories compiled from letters she and her husband Paul wrote to friends and family during those years. At the beginning of her marriage, Julia confesses she knew little about cooking. Her first meal as a bride – brains simmered in red wine – was a flop. But her arrival in France, and in particular, a lunch of sole meunière in Rouen, introduced her to the "revelation" of French home cooking, which would shape the rest of her life. Child describes classes at Le Cordon Bleu cooking school; her first attempts at French cuisine in their eccentric Parisian kitchen; beginning a cooking school, L'École des Gourmettes, with friends; and finally, her quest to compile a book of "French recipes for American cooks," which would come to be known as *Mastering the Art of French Cooking*.

After reading the first chapter of Robert Farrar Capon's ***The Supper of the Lamb: A Culinary Reflection*** (Modern Library, 1969), you will never cut an onion the same way again. The work is an unconventional, yet utterly delightful cookbook by an Episcopal priest with a passion for theology and cooking. Capon refreshes the reader's appreciation for good food by challenging our tendency to ignore the material world in our quest for sanctity. "Food is the daily sacrament of unnecessary goodness," he writes, "ordained for a continual remembrance that the world will always be more delicious than it is useful."

Jane Sloan Peters is a doctoral candidate in historical theology at Marquette University. She lives in New York City with her husband, son, and cookbooks.

The book is loosely structured around a single recipe: Lamb for Eight Persons Four Times. There are many asides, in which Capon offers charmingly haphazard lectures on the glories of butter, essential kitchen gadgets, and how to throw a proper dinner party. He extols the turnip, "one of the lordliest vegetables in the world," and the mushroom, "Ah! It is the proof of creation ex nihilo." For all the whimsy, the theological reflection is precise and edifying. Before an appendix of bona fide recipes, Capon closes with a chapter for heartburn remedies. But there are two kinds of heartburn – one may be cured by a baking soda solution, the other (the "Higher Heartburn") is a longing for heaven that can only be cured when we reach it. Yet "heaven does not run *from* the world but *through* it," and so we offer God every good thing as we prepare to meet him. *The Supper of the Lamb* will leave the reader laughing, and longing for both a good meal and the Everlasting Banquet.

Be prepared to get flour all over Daniel Leader's **Local Breads** (W. W. Norton, 2007). Leader, the founder and CEO of Bread Alone Bakery, visited bakers across Europe and compiled recipes from their centuries-old methods of maintaining sourdough starters and producing artisan breads. "Many of the recipes in this book come directly from a particular baker," Leader writes, and he chronicles his encounters with characters like Stefano Galletti, a Florentine baker who proudly explains *doppio zero* flour, or Clemens Walch, who specializes in rye breads at his bakery in the Austrian Alps. The cookbook devotes three chapters to French breads – from the classic baguette and *pain de campagne* to darker breads from Auvergne; four chapters to Italian breads – featuring ciabatta, semolina sourdough, and a few pizza recipes; and two chapters to rye recipes from Germany, the Czech Republic, and Poland. Best of all, Leader gives detailed instructions on how to cultivate European starters and sourdoughs at home: the surest way to approximate bakery-bought bread in one's own kitchen. There are also helpful troubleshooting sections for when loaves fall flat or starters lose their zip. *Local Breads* is perfect for the aspiring bread baker willing to invest the time to get it right.

Maria von Trapp's **Around the Year with the von Trapp Family** (Sophia, 2018) has been reprinted for the first time since its original publication in 1955, giving a new generation of families the opportunity to celebrate religious feasts with prayers, songs, and recipes from the von Trapp family. Maria saw the Catholic liturgical year as a "school of living" that enlarges our understanding of "feast" by teaching us to celebrate in the Lord. Other Christian denominations will appreciate, too, her careful selection of traditions for Christmas, Lent, Easter, and occasions such as birthdays, graduations, and weddings. "It is the feast that helps to keep the family together," she writes. Recipes – mostly sweets and breads – include sacher torte, a rich chocolate cake, for birthdays; unleavened bread for Lent; and lebkuchen, an Austrian gingerbread, for Advent. ➘

The Greatest Thing Since Sliced Bread

Meditations of a Meat-Eater

SARAH RUDEN

IN EVERY GOSPEL version of the Last Supper, and in Paul's version too, a strange textual sleight of hand occurs – the kind of thing that can easily go unnoticed in such a beloved and foundational story. From the Passover, the feast of unleavened bread, the feast itself disappears. The sacrificial animal is so essential that the same term, *pascha*, can mean both the roasted carcass and the whole meal. In the Greek of the New Testament, the disciples ask Jesus where they should prepare "to eat the Passover" – to eat the *pascha*. The traditional food is prepared, and they eat, but when it comes down to it, no one is shown eating the all-important course. In Luke, Jesus seems to refrain, saying that, despite his longing, he will not "eat this Passover until it is fulfilled in the Kingdom of God" (22:15–16). Explicitly shared around, under Jesus' command to consume them, are of course only wine, the ordinary drink at meals, and bread.

But chapter 12 of Exodus contains the instructions for a truly celebratory meal. Animal protein was a luxury in the ancient world, and for the Passover, scripture went out of its way to specify what would have been a true treat: a one-year-old sheep or goat (nearly full-grown, that is, but still tender) for each house. The community is to gather and slaughter the animals in public – which means that, once there was a Temple, its officials did not take a share of the meat. The carcasses could be shared between small households, but the process could leave no individual out. ("You should apportion the animal according to what each person will eat.")

Each group of celebrants had to finish off their meat in one evening and burn any leftovers. This commemorates the hasty meal before departure from Egypt, but another impression must have been a lavish and confident banquet: "We're eating meat here, and we're not putting aside any of it."

In the relatively cosmopolitan and urbanized culture of the first century AD, sacred law had found a number of workarounds. The Passover festival couldn't in fact look like the homey rites prescribed in Exodus, but involved

Sarah Ruden is a poet, translator, essayist, and popularizer of biblical linguistics.

massive pilgrimage to Jerusalem, and all the necessary commerce was in place. But I intuit that the sustained gift-giving character of the Passover – at early harvest time God renders the people prosperous enough to eat meat, and household members share it among themselves with no subtractions for officialdom – helped send Jesus over the edge on this occasion, into his attack on people doing business in the Temple courtyard. The Temple, gigantic and colorful (Herod the Great, allied with pagan rulers, had expanded and refurbished it), must have looked to him the way a pseudo-sacred theme park would look today, with ticket booths, turnstiles, and hotdog stands.

After the public commotion comes the private banquet, the Last Supper. Jesus and his followers, observant Jews, would have had a Passover sacrifice to make short work of, but readers don't observe them eating it. Jesus himself is about to become the "Lamb of God," and "Christ our Passover" who is "sacrificed for us," the flesh made (in him, briefly) lifeless to feed ongoing life. But the only food Jesus is shown dividing and distributing is bread.

This would be the unleavened bread of the Passover, bread that didn't have time to rise before the flight from Egypt. The "bread of haste" isn't *supposed* to be savory. Matzo is hard and flat, with no extra salt or other spice like a cracker's, and without even the flavor that yeast adds. Unleavened bread is slavery's plain food made abnormally plainer.

So why does bread come forward at the Last Supper, pushing aside precious meat at the very moment when the dispersion of meat is being enacted festively and is about to become a new symbolic prescription? Jesus' body is given to all of humankind, miraculously bountiful like the few fish multiplied over the five thousand – a rich, eternal feast. Why is the Passover meat invisible? Why does the Lord's Prayer ask only for daily bread, when so much more was provided?

IN SOUTH AFRICA, against a background of destitution and subsistence struggle, it is striking how strong the human drive to share meat can be. This struck me again and again during the nearly ten years I lived in Cape Town. As a guest at a club for retired housecleaners in a "black township" outside Cape Town, I was once given a heaping plate of chicken and struggled to do it justice, anxious to be gracious but concerned that I was taking leftovers away from several grandchildren waiting in the women's homes.

Events such as traditional weddings and funerals, which demand the inclusion of the whole community ("If they know about it, they will come"), can devastate households. When interviewed for a documentary, a white clergyman reported that when he visits the African bereaved to ask what they have to live on after the death of their only wage-earner, he may well hear that they have nothing at all: they cleaned out the bank account to feast their near and distant neighbors.

Against this background, the habits of middle-class and wealthy whites who made up my social circle are troubling and understandable at the same time. Their feasts are lavish but exclusive. One year, the established host in my circle of professors and aid workers bought a

Photographs courtesy of Rachel Korinek

shallow second-hand metal cart, about ten feet long and six feet wide, to convert into a grill big enough for the Afrikaner *braii* or barbecue. I counted – I think – ten kinds of meat crowding over the coals, including cuts too fatty for me ever to have seen in the States.

But we were discreet, in ways unspoken even among ourselves. I can recall only one African servant ever being present around the time of a party. This left the hosts a great deal to do, but they did it cheerfully; other white hosts I knew had some help with food preparation or clean-up, but that help was enclosed and muffled in a distant kitchen.

In my own home, I had adjusted to the overarching rule that servants were to be connected as loosely as practical to the food. Maids everywhere seemed to cooperate, withdrawing quickly with the food they were given, and professing themselves helpless in the kitchen except for cleaning. To me as a Quaker, at least nominally pledged to equality, the rule was shocking on its face, and I started out breaking it, serving lunch to my maid and foreign lodger at the same time and sitting down with them. But the maid was plainly distressed, so I took to filling a plate for her and letting her take it out to the porch alone.

After two short stints back in the States, and ending another long one in South Africa, I found I had been in an ethically impossible place. Wishing to give a farewell treat to a second beloved maid, Lucy, I asked a neighbor for whom Lucy also worked to drive us both around the scenic Cape Peninsula and join us in a restaurant meal. The neighbor agreed, but only on the condition that Lucy not have the whole day off but show up early to clean her bathroom. I was given to understand, however, that the ultimate imposition on my part, the hard evidence that I was out to spoil the help, was Lucy's plate of soigné fish.

I'm still not sure exactly what I was up to: whether or not I was being self-righteous and impractical; whether I risked straining Lucy's relationship with a household that valued her and provided her long-term security. What was the point, after all, of rubbing it in that Africans' daily food, if they were lucky, was *mieliepap*, a thick porridge of cornmeal?

The cornmeal was one of the basic grocery items angrily named by my students when I suggested that a few rand from their modest stipends should purchase used books. It was the square of yellow powder in thin plastic bags doled out to roadside work-seekers by "Mr. Inasmuch" (so named by us because in the King James Version, Matthew 25:40 reads, "Inasmuch as ye have done it unto one of the least of these my brethren, ye have done it unto me"), a retired prison chaplain whose good works the Cape Western Monthly Meeting of the Religious Society of Friends helps to fund.

We visited him at his tasks one morning. The peanut butter sandwiches and soup were for now, he explained to us as he doled, but the *mielie-meel* was to take home in case not enough foremen came by in their trucks to recruit for building sites, which meant that for

Photograph courtesy of rawpixel

going home empty-handed to their hungry families. I can't find a trace of the old man's operations on the internet, and I know from Quakers' own long-ago ventures that food interventions entail mammoth difficulties. The first among these is to find someone who gets up at five every weekday morning, propelled by her clients' condition, as if their hunger were her hunger, their bodies her body.

IT SHOULDN'T HAVE taken these observations to awaken me to the steepness of the human fall, from the meat of hunting and pastoralism to the daily toil of farming ("By the sweat of your face you shall eat bread," Adam is told when he and Eve are expelled from the garden), on plots that from the first needed defense against outsiders and later were fragmented to keep up with increasing populations. Still later, these plots were consolidated by overweening powers that doled grain to the masses of uprooted urban poor, to slaves on estates, and to common soldiers. Throughout the Roman Empire around the time of Jesus, these were the main ways in which food was distributed, with concessions in Jewish and early Christian culture in the name of charity: the poor deserved to eat not because they were useful to keep alive or potentially dangerous if starving, but because, as fellow human beings, they were worthy to enjoy an imitation of God's compassionate care.

That's the history I know now. But as I knew early on from my parents – Great Depression children who lost family farms (one quickly, one slowly) and grew up to be biologists (my father's research stretched from crop pests to demographics) and environmental activists – we've fallen a lot farther than the

some work-seekers there would be nothing to feed the family in the evening. The readiest recourse then was to ask the wife to go sell herself for the five rand that would buy a bag of meal this size.

I guessed that it weighed two or three pounds; in many cases it would need to feed ten people for twenty-four hours or over a weekend. "You pure Quakers won't like this," Mr. Inasmuch told us, "but sometimes I have to give them a little *klop*. They tell me, 'You can't hit me – you're not my father.' I say, 'I *am* your father, because I feed you.'" Right in front of us, when one of the men moved away from the wall with which he was supposed to stay in contact, Mr. Inasmuch slapped him upside the head.

But we revered this old man, who was feeding thousands of people alone. When we offered more money, he said that what he needed was help; he was becoming too frail to do the work alone. Could someone join him, with a view to taking over eventually? There were no takers. "Casual" workers who wait on the bleak sandy roadsides between warehouses outside Cape Town probably now lack any source of daily grain, any assurance of not

Romans. Now, as millions of people face the end of the earth's capacity to provide for them, we may simply be stuck before the spectacles of our era's main horrors: the mass violence of redistribution, and the mass violence of inequality.

My parents had considerable insight into where they themselves had, on a rare chance, landed: as American middle-class mid-century professionals, beneficiaries of the GI Bill. They knew they were on a precipice of meat, towering over grain. They sneered at new cars, Disney vacations and films, even air conditioning; but a whole cow or half a cow from a local farm was always wrapped, cut upon succulent cut, in white paper parcels in the chest freezer in our mud room. The freezer was so big that as a child I teetered on its edge, the lid resting on my back, while I rummaged for whatever my mother in the kitchen required. When money was low – and when it wasn't – my father hunted. I have a photo of him grinning, a freshly killed wild turkey in his arms, the tail spread out and reaching down to his knees.

I had a constantly guided instinct about the inferiority of starchy foods to anything that could be killed, grown in a garden (ours were large and ambitious) or orchard (we had a tiny one), or found in forests or meadows. Since I'd been taught to forage and advised how good sorrel and certain berries were, I did try wild grains when I found them, but they were repellent: unpleasantly chewy, with husks and tails that caught in your throat, and with none of the natural flavors encouraging you to eat things you can digest to your body's profit.

My father flaunted in the kitchen while my mother was away at nature-study camp. He fed me – a skinny kid – one steak and then quietly, as if moving in on a pheasant, asked whether I'd like another; he fried it up and delivered it, overstretching the plate, before I could reconsider my yes. He beamed and hovered as I ate it. Steak was glorious, and just about effortless to cook. His oatmeal, in contrast, stuck to the saucepan, and reared in the cereal bowl in slimy, rubbery slabs. You couldn't swallow his experimental homemade bread, made without salt. (Salt in the ancient world usually depended on distant trade. What if you couldn't get salt to make your bread palatable?) Grains cost practically nothing to ruin, and were almost worthless as a success.

I have the preference for meat literally in my big bones, so that I've never subscribed to the vegetarianism and veganism popular among my fellow Quakers. I can admire some of the big principles – my own mother swore off beef for a number of years, as the worst offender in global dietary inequality – but I myself have embraced only humane standards for keeping and harvesting the animals that feed us.

> We may be stuck before our era's main horrors: the mass violence of redistribution, and the mass violence of inequality.

Any milk or beef I buy myself is from certified 100 percent grass-fed animals. The eggs in my kitchen are free-range, not just cage-free, and like my bacon, pork, and ham, they have the rare, all-important CERTIFIED HUMANE RAISED AND HANDLED logo that *Consumer Reports* says means *actual* decent treatment of the animals. I lavish more time, trouble, and money by far on my protein than on any other part of my consumer acquisitions, in which I'm generally a pretty conscientious, simply living Quaker: with no car, no smartphone, no

Highly processed carbs can seem like an apt representation of human fallenness; they are a bad artifice; they are what we, not God, created, and look what our creations, our confections, do. They spoil the character of *animals*. Sugar lumps make horses mean, I learned when I obtained a pony. The garbage dumps around Cape Town, with their burger bun scraps and half-empty bags of cheese curls, transform well-adapted baboons into obese hoodlums.

I myself was like Edmund when I emerged from my own home, which offered hardly any junk food, and tasted potato chips and chocolate at Brownie camp or sleepovers. I tasted, and was downgraded as a human being; no longer interested in the other girls who were ready to play games with me or tell me secrets and hear mine, or in the adults who wanted to guide and entertain me: I just wanted to maneuver to get more of that stuff in my mouth, and they could see what I was up to, and I could feel their contempt.

"Good food" was what constructed me as a human being, and "garbage" was the drug that tore me apart, did away with my proper, sustainable energy and restraint. "Good food" was the eggs and roast beef from the neighbors, and the venison from my father's rifle, and our homegrown fruits and vegetables, all of which my parents put in front of me. Other people, elsewhere, would let me suck on sweets or stuff my gullet with chips, but that was because they didn't care how I turned out.

For this reason, I wouldn't become a grain-sustained vegan in solidarity with the people who have to live on starch, any more than I would live on the street in solidarity with the homeless. But I know I stand on an edge, where my deliberations and choices are absurd. They are very far from the innocent, hard-won prosperity of my parents, or their sincere hope

cable or even broadcast TV, no designer labels or fast fashion, make-up applied only once or twice a year.

My health is one reason for the things I buy in the health-food store down the road. My palate is another. Another is the fear of appearing before the throne of God and seeing beside it thousands of animals who spent their lives immobilized, crippled, bald, decaying already, uncomprehending as to what landed them in this fattening hell – and now they know that it was my idiot hand reaching for a gas station hot dog or a pale slab of lunch meat.

But my deepest motivation is my lifelong sense that the jazzed-up carbohydrates of globalized food equality are not just unhealthy but penetratingly dehumanizing. In *The Lion, the Witch, and the Wardrobe*, the Snow Queen corrupts Edmund with a single helping of Turkish delight; he tastes that gelatinous sugared treat, and betrays his siblings and friends and all the forces of good on the mere promise of more. He's soon locked up and chewing on dry crusts – a discarded slave, to be rescued only by the lion Aslan's death on a table of ceremonial torture; that is, by the actual meat-eater's gift of his own flesh.

that the Green Revolution, lower birthrates, renewable energy, or the combined ingenuity of themselves and all their fellow wonks could someday bring us all to the same feast. I know what I've seen, in Africa where I've lived. I can't conceive how we will all be able to share anything worth sharing.

WHERE IS THERE ANY future? While learning German, I came across a short story that shows one particularly dark chasm down which humanity doesn't have to fall. The story is by Wolfgang Borchert and called *"Das Brot"* or "The Bread." It is set soon after World War II, at a time of stringent food rationing. A woman is roused by a noise at night, to find her husband in the kitchen in front of the loaf, the knife, and fresh crumbs. She refrains from calling him on his theft of her share of food, and the two agree that they both rose in response to noise from the rain gutter. They retire to bed again, but when the man thinks his wife is asleep, she can hear him chewing. Even then, some powerful inner force keeps her silent.

Machine-slicing was patented in 1928, after which commercial bread could easily be divided into equal portions. But apparently sliced bread wasn't available, or common, in postwar Germany. The loaf in the story is cut by hand, so that one of the people entitled to share it can cut in secret, take more than is fair or agreed on, and leave his partner unsure whether it now looks smaller. If this goes on, the woman could starve; all her dutiful contributions to the household's survival might prove inadequate to save her own life. What do marriage and the family mean, in that case? The story points quietly, through the deep meaning of an ordinary object, to the horror of society's breakdown under fascism and war.

There are many words for wrongdoing in Hebrew, Greek, and Latin, but as a translator of sacred literature, I've long since made the rough division in my mind between those terms directed at a childlike sensibility and those directed at an adult one. Children are admonished not to go where they shouldn't, damage or dirty what is clean or whole (including their own bodies), say or touch forbidden things; the language of purity and obedience covers one large moral domain. But fiercer words are applied to people who have power and property, insight and skill, and yet take the community apart by lying to them- selves and others about how deadly a disadvan- tage to someone else might result from their own trivial advantage. Their misdeeds are those of conscious disproportion. They take and give bribes, appropriate land, abuse and exploit those they are supposed to protect, cheat their underlings, annihilate those who merely annoy or embarrass them.

> Highly processed carbs can seem like an apt representation of human fallenness.

In the traditional English liturgy, we speak of both our "sins" and "iniquities." The word "sin" is thought to originate from the Latin used in court proceedings and means some- thing like "That's him!" – authority points at and blames someone already under its power. "Iniquity," on the other hand, means "uneven- ness" or "inequality." At first this might not seem to be on par with the designation of, say, a murderer. But think of the way, in circles where the word "sinful" is a quaint joke, people may still hiss the word "iniquitous," with a hatred for the cruelty and indifference of power. It's relatively easy to bring ordinary criminals under control. It's practically impossible to get at systematic, privileged offenders.

Much the same sentiments prevailed in antiquity, and may help explain why we sanctify bread. The edible parts of a single carcass are hard to divide precisely; the breaking points are tricky, the quality of the meat varies. Not so with homogeneous, storable, and malleable grains, which lend themselves to standardization.

Presumably, the bread in the story *"Das Brot"* has been weighed by the authorities – so many grams for each person in each household – that is, it has been fairly divided, to a point. There were such dispensations in Old Testament times, as evidenced by stern Bible verses about "just" weights and measures (Lev. 19:36, Deut. 25:15). Under Roman regimes, a whole class of magistrates, the aediles, patrolled the accuracy of scales and the size of containers in marketplaces.

> The meat of the feast disappears because the bread itself becomes the meat.

But bread itself could come pre-divided as well. The Roman *panis quadratus* (examples ossified by volcanic heat have turned up in the Pompeii excavations) was made in a pie-shaped mold, so that sections could be torn off like pizza slices; the indentations rendered them all the same size. In Palestine, square bread molds could include perforations, for guided tearing in the manner of modern payment slips. Perhaps the bread Jesus "broke" looked like some of the matzo mass-produced today, or like our ordinary crackers, which split into equal portions under just a little pressure.

The disciples had been feasting, enjoying a whole roasted lamb; they wouldn't have been fretting about who got how much bread. Thus the remarkable thing is the focus of the scene, as far as physical action goes, on the approximation of an act constantly repeated in thousands of places throughout the Roman Empire: an authority handed out some form of grain food in some kind of standard equal measure.

Unequal distribution must have happened from time to time, but in general it was considered just too outrageous to cheat of minimal food indigent citizens or "clients," children, hired laborers, or slaves; some of which people you could buy and sell, work or torture to death, rape at however young an age. Functional social bonds couldn't outlast any sustained toying with the bare subsistence of dependents; they were all owed their share. Roman satirists show how unequal a rich man's banquet could be: he sat people according to rank, lowest to highest (the Bible indicates this happened among Jews during the time of Christ); he could serve them separate, highly unequal hors d'oeuvres and entrées. But if his overseer in the back room dealt unfairly from the same jar of grain or the same loaf, it would have been too much. The lunch lady can't give one pupil an extra scoop of macaroni. The Salvation Army officer can't give one homeless person two helpings of soup. The institution would not hold.

Minimally equal apportioning by geometry in various forms is in fact at the (sometimes literal) foundations of civilization: plots of subsistence land, dwellings, rooms within dwellings, and tenements in a multi-story building (already existing in antiquity); tables and food containers in a marketplace; the grid of streets; the seats in an arena, the enrollment list, the military march and the fight in formation; the choral dance, the religious procession, the seats in a classroom, and on and on. Certain kinds of inequality are unbearable; they are not allowed; they would threaten to

Photographs courtesy of Rachel Korinek

bring down the rage of God in the form of society undoing itself. In fact, Mr. Inasmuch, and all the other people in Africa whom I saw giving lifesaving aid, were strict about physical space: lines, locked boxes, storerooms, and cars; their work would come apart if these limits weren't maintained.

So if custom had anything to do with Jesus' handing around of Passover bread, it wouldn't have mattered whether one disciple was the sometimes unidentified one "he loved"; or Peter, the one he had designated to found his church; or the three who had witnessed the transfiguration manifesting that he was the son of God; or Judas, the one who was going to betray him this very evening. He dealt justly, according to the law he came to fulfill and not to destroy. In the Gospels and Paul, there he is, the Messiah, holding this precise, chore-like line. But he himself is the actual human embodiment of the Passover sacrifice: the best for everyone, endless, self-multiplying, unconfined joy for human nature.

HOW CAN THE PARADOX work? During the time I spent in divinity school, which was less than ten years ago, controversy over transubstantiation was ongoing with a vengeance. A student raged that, if crumbs were carelessly dropped during communion at chapel, she would feel compelled to lick them off the floor. Another student was bumped off an ordination track when she refused to venerate the consecrated host as embodying Jesus. At moments, I felt almost as if I was living in England under the Test Act of 1673, which made denying transubstantiation the sole indispensable qualification for serving the state in any capacity. Bread vs. meat in the Lord's Supper must strike a deep nerve, to become language and tradition of such consequence.

If I reject theological formulae, it's because they don't go far enough. God's love and providence are real enough to me that I believe Jesus' body and the bread he distributed were in some vital sense the same already, and in some vital sense always will be – and not mysteriously, but manifestly and commonsensically.

This all needs to be about love, I'm convinced. The meat of the feast disappears because the bread itself becomes the meat, with both the commonality of a staple and the richness of a nourishing treat. Love defies even the laws of physics, multiplying through its own power. The children's song about love as the magic penny ("Hold it tight, and you won't have any / Lend it, spend it, and you'll have so many . . .") endures in the adult mind as true.

There's no understandable reason for joyful selflessness to go on and on flourishing. But somehow it does. ⇥

Why Yemen Starves

The Making of a Modern Famine

DANIEL LARISON

A child at a shelter for displaced persons in Ibb, Yemen, August 2018

MODERN FAMINE is almost never the result of a lack of food. This may seem strange; for almost all of human history, people have starved when crops fail or wars deplete food supplies. No longer. Today, famines are man-made. Nor do they happen by accidents of omission. Often political leaders choose to inflict this punishment on a group of people whose lives they consider expendable. To create a famine in the twenty-first century requires an extraordinary amount of organized effort. It is something that some people do to others to achieve their political goals. As such, it ranks as another type of mass atrocity and a crime against humanity. One such crime against humanity is taking place today in Yemen.

Yemen has been hard hit after more than four years of war. As Alex de Waal tells us in his valuable history of modern famine, *Mass*

Daniel Larison is a senior editor at the American Conservative, *where he also blogs. He holds a PhD in history from the University of Chicago, and lives in Lancaster, Pennsylvania.*

Photograph by Mariman El-Mofty. Used by permission.

Starvation, "acts of commission – political decisions – are needed to turn a disaster into mass starvation." Indeed, Yemen's famine is largely a product of economic blockade and other policy decisions made by the Saudi-backed, internationally-recognized government of Yemen led by President Hadi. Hadi was the successor to Ali Abdullah Saleh, who had ruled Yemen for over thirty years when he was forced out following protests in 2011. Hadi was himself then ousted by Ansar Allah, also known as the Houthis, in a September 2014 coup. In the spring of 2015, a coalition of Arab governments led by Saudi Arabia and the United Arab Emirates, and supported by the United States, launched a military intervention to reinstate Hadi and expel the Houthis from the capital. Saleh and the Houthis formed an alliance of convenience that collapsed last year when the Houthis fell out with Saleh and killed him. Today, the coalition is no closer to achieving its goals, but the civilian population of Yemen has been thrown into the abyss.

Between the damage done to the country's infrastructure from Saudi coalition bombing, the sea and air blockade maintained by the US-backed Saudi coalition, the relocation of the central bank to Aden, the devaluation of Yemen's currency, and more than two years of failing to pay civil servants their salaries, Yemen's economy has virtually collapsed. That has meant deepening poverty for most Yemenis. As many as fifteen million – more than half the population of the entire country – are so food insecure that they are at risk of starvation. There may be food in the marketplaces in Yemen, but it has become prohibitively expensive for a population impoverished by conflict and inflation. And the economic war being waged on the civilian population is causing far more deaths from preventable causes than bombing and shelling.

Save the Children estimates that at least eighty-five thousand children have starved to death since 2015.

CHILDREN ARE USUALLY among the most vulnerable to the ravages of famine, especially because malnutrition puts them at greater risk of dying from disease. The sad story of Amal Hussain, a seven-year-old Yemeni girl, is representative of the plight of millions of children in that war. The *New York Times* first reported on her condition in late October 2018, and the story was accompanied by a haunting photograph of Amal's frail body wasted by extreme hunger and diarrhea. Within a few days of the report, Amal had died. Amal's family had lived like refugees in their own country since their home was destroyed by a Saudi coalition airstrike three years earlier. It was in the camps for the internally displaced that she slowly wasted away. Millions of Yemeni children are just as severely malnourished, and their families are just as poor. Even the children that don't perish from hunger and disease have had their development stunted and their lives permanently scarred by the experience of living through war and famine.

Just as famine has political causes, it can have a political remedy. Unfortunately, these atrocious famines have not generated the attention or interest worldwide that other mass atrocities receive. The countries affected by famine are not covered in the news media very often. When there is coverage, it seems to have little or no effect on policymakers and the public. There is a real danger of famine making a comeback in many countries where outside governments are either complicit in causing mass starvation or have no interest in staving off disaster. After nearly succeeding in eliminating famine entirely, the world seems mostly oblivious to its horrendous return. ➤

The Ground of Hospitality

NORMAN WIRZBA

THE CHRISTIAN CHURCH has multiple saints associated with gardening, farming, and food. There's Saint Isadore, Saint Urban, Saint Fiacre, Saint Patrick, and, of course, Saint Francis of Assisi. I especially like Saint Phocas, who farmed in the region of Sinope, a fairly remote Turkish peninsula jutting into the Black Sea. During his life he aided Christians being persecuted by the emperor Diocletian (284–305). He regularly opened his home to strangers and travelers, and he fed the poor with food he had himself grown. Phocas was well known across the country for being a man of great charity and virtue.

It was his reputation for hospitality that got him into trouble. Scholars believe that sometime around AD 303 Diocletian singled him out as one Christian especially deserving of death. The emperor promptly dispatched two soldiers to do the job. They had an enormous distance to cover, much of it through treacherous and hostile terrain. As they approached Sinope, they came upon Phocas's farm. They explained their mission to him. Phocas insisted that they spend the night at his home. He assured them that he could direct them to the man on the following day. Meanwhile, they needed a good meal and a good night's rest, which he provided.

> While the soldiers slept, Phocas went to his garden to dig a grave.

While the soldiers slept, Phocas went to his garden to dig a grave.

The next morning, having enjoyed a good breakfast, the soldiers asked for directions. Phocas told them they need look no farther, since he was the man. The soldiers were astonished, perhaps even embarrassed, having just enjoyed such extraordinary hospitality. But Phocas insisted. If they failed to kill him, the emperor would likely kill them. And so they went to the garden, where the soldiers cut off his head and buried him in his grave.

I have no doubt that Christian training played a major role in Phocas's development as a man widely known for his kindness, generosity, and hospitality. But I also believe that his work in the garden contributed directly to his hospitable manner. By giving himself to the ground he learned in the most visceral manner that the ground is constantly giving in return. The ground isn't simply a stage on which people do their thing. It is, rather, the miraculous matrix that cradles, supports, and feeds life.

To work in a garden is to be surrounded by the mysteries of germination, growth, and decay, and it is to be overwhelmed by the gifts of raspberries, tomatoes, and onions that surprise us with their fragrance and taste. But

Artwork by Ángel Bracho

it isn't all pleasantries. To garden is also to be frustrated by the disease and death that are beyond one's control and power. Where did this blight come from? Why won't this seed germinate? A late frost again? The temptation is always to give up and walk away. But that isn't really a viable option. If people are to eat, they must eventually return to the ground.

Saint Phocas understood that if the land gave itself to him, then he must give himself to the land in return. He did this daily, in the actions of cultivating, seeding, watering, weeding, pruning, and harvesting. He didn't simply take what the land provided. He invested his time and skill, his devotion and energy. At the end, he even gave his body to the ground. I like to think that it pleased him to know that his flesh and blood would nourish the ground itself, feeding the life below that feeds the life above.

Gardening is one of the most vital practices for teaching people the art of creaturely life. With this art people are asked to slow down and calibrate their desires to meet the needs and potential of the plants and animals under their care. Gardeners are invited to learn patience and to develop the sort of sympathy in which personal flourishing becomes tied to the flourishing of the many creatures that nurture them. A garden, we might say, is a living laboratory in which we have the chance to grow into nurturers, protectors, and celebrators of life. This, I believe, is why the first command given to the first human being was to come alongside God the Gardener and "till and keep" the Garden of Paradise (Gen. 2:15). Gardening is hard and frustrating work, but it is not a punishment. To garden well – in the skillful modes of attention, patience, sensitivity, vigilance, and responsiveness – is to participate

Ángel Bracho, *The Wheat Ear*

in the way God gardens the world.

Among contemporary writers, few have understood and articulated these insights as well as Wendell Berry. Whether in the form of poetry, story, or essay, Berry has argued that apart from a people's commitment to repair and nurture particular places and communities, the world comes to ruin. His call to "return to the land" is not the expression of some romantic yearning to relocate urbanites within an agrarian arcadia that never existed. The issue is not relocation, but the development of the sympathies and skills that make for an enduring, responsible, and beautiful livelihood. One doesn't need a farm to do that. All one needs is a place within which to learn to exercise care and commitment. He knows it won't be easy, especially in cultures characterized by

---

Norman Wirzba is Gilbert T. Rowe Distinguished Professor of Theology at Duke University, and the author of several books on topics from environmental ethics to sustainable agriculture.

speed, rootlessness, and a spectator approach to life.

From an agrarian point of view, one of humanity's most important postures is looking down. Though plenty of spiritualities encourage people to look up and away to a better world beyond the blue, looking away causes us to forget that in fact the ground beneath our feet nurtures us. Scripture made the point inescapable (Gen. 2:7, 3:19): to say the word human (*adam*) is to be reminded of the ground (*adamah*) from which we come, by which we are fed, and to which, upon death, we return. To ignore the soil or, even worse, to despise it, is to cut oneself off from the love of God and the power of life that circulates through it.

As people have moved out of agrarian ways of life, soil has mostly disappeared from their imaginations. If it registers at all, it is often as "dirt" – the substance to avoid because it makes one "dirty." This is a tragedy, because soil is indescribably complex and miraculous in its ability to create the conditions for life. In his early essay "A Native Hill," Berry described it this way:

> The most exemplary nature is that of the topsoil. It is very Christ-like in its passivity and beneficence, and in the penetrating energy that issues out of its peaceableness. It increases by experience, by the passage of seasons over it, growth rising out of it and returning to it, not by ambition or aggressiveness. It is enriched by all things that die and enter into it. It keeps the past, not as history or memory, but as richness, new possibility. Its fertility is always building up out of death into promise. Death is the bridge or the tunnel by which its past enters its future.

It would be a mistake to dismiss this characterization as a poetic flight of speech. Hans Jenny, one of the great soil scientists of the twentieth century, noted that sixty years of study only reinforced his realization that soil is fundamentally a mystery. The border between life and death, the biotic and the abiotic, is nearly impossible to draw. Soil is constantly receiving massive quantities of plant and animal corpses, and so should be a stinking mass of death. But it isn't. Somehow death, by circulating through soil, is transformed into the fertility and fecundity of life.

Soil, we could say, is the first earthly site of hospitality, because it makes room for death, welcomes and receives it, so that new life will germinate and grow. The more primordial power of hospitality, however, is God's. For good reason, the Garden of Eden story presents God as the one who creates by kissing soil, breathing into it the life that is you and me and all the plants and animals. In this gesture, God communicates that the divine nature is never to be far away or aloof. God is near, and stays intimately close as the breath within our own breath and as animate soil that circulates throughout all our eating. God's creating, creative power is a hospitable power that constantly *makes room* for everything else to be and to flourish. God is the primordial host who prepares the beautiful, fragrant, and delectable feast at which all creatures are fed and find their true home.

In his book *Life in the Soil*, the biologist James Nardi takes readers on a fascinating journey into soil. By following the routes of roots, he notes that after just four months, a single rye plant will send down 15 million roots totaling 380 miles. These roots make surface

> ## We must become hospitable to the soil that is hospitable to us.

contact with an area of approximately 2500 square feet. If one adds to the equation the innumerable, miniscule hairs that attach to roots, then the length of the overall root system extends to 7000 miles in length and 7000 square feet in surface area.

Above ground a plant may appear to be a solitary, self-standing thing. But the roots reveal a different story. Plants crave contact and (chemical) communication. To be healthy, they need a dense network of nurturing relationships. A healthy plant, however, doesn't simply take from the soil and all the microbes alive within it. The plant receives sunshine and transforms its energy into food, especially sugars, that it sends down through the roots to feed the fungi and other microscopic creatures that make their home near the roots. The more the roots grow, the more hospitable the soil becomes, further aiding the fertility of life. The vitality and vigor of plants, not to mention the tastiness of their fruit, depend on maximizing the flow of hospitality that circulates through sunshine, stems, roots, and soil. The destruction of life begins with the erosion, denuding, and poisoning of soil.

Agrarians believe that few tasks are more fundamental than for people to become hospitable to the soil that is hospitable to them. The work of making room for others, noting their need and potential, and committing to care for them, is the indispensable work. It is here, in the giving and receiving of nurture, that we learn the meaning and the point of life. If you want to experience life's abundance and potential joy, give yourself away. This is what the gospel teaches. It is what God has been doing since the beginning. It is what the soil witnesses to every day. ⤳

from "ENRICHING THE EARTH"
Wendell Berry

To enrich the earth I have sowed clover and grass
to grow and die. I have plowed in the seeds
of winter grains and of various legumes,
their growth to be plowed in to enrich the earth.
I have stirred into the ground the offal
and the decay of the growth of past season
and so mended the earth and made its yield increase.
All this serves the dark. I am slowly falling
into the fund of things. And yet to serve the earth,
not knowing what I serve, gives a wideness
and a delight to the air, and my days
do not wholly pass. It is the mind's service,
for when the will fails so do the hands
and one lives at the expense of life.
After death, willing or not, the body serves,
entering the earth. And so what was heaviest
and most mute is at last raised up into song.

Source: Wendell Berry, *Farming: A Hand Book* (Counterpoint, 2011).

Beating
the Big Dry

How an Australian cattle farm is
fighting drought by reviving
ancient landscapes

S ince the year 2000, historic droughts and flooding rains have hit Australia's farmers hard. Pastures have withered in affected areas, forcing farmers to shoot starving sheep and cattle. Could there be a different way to practice agriculture that helps the land flourish even amidst climate change? *Plough*'s Chris Voll talked to Johannes Meier, who runs the Danthonia Bruderhof's farm in New South Wales.

Johannes Meier

Plough: In May 2019, Danthonia Bruderhof celebrates twenty years since its founding. Tell us about the community's ongoing discovery of how best to farm this land.

Johannes Meier: In 1999, the Bruderhof purchased two neighboring farms in the Northern Tablelands district of New South Wales, Australia. It's an area known for its agriculture and is deemed reasonably rain-safe, at least on paper. Two families and a few singles from the United States packed their bags and boarded a Qantas flight – and so began the Danthonia Bruderhof community.

We arrived full of enthusiasm – and inexperience! To help us, for the first year we retained the farm manager from Danthonia's previous owner before taking on the farm ourselves. Like most farms in this district, Danthonia had run a mixed operation, grazing beef cattle and merino sheep, and cultivating standard crops: grains in winter, beans, sorghum, corn and sunflowers in summer. We continued planting crops for that first year, and it went well. But we then decided to focus just on livestock.

In those beginning years, we discovered the hard way how costly it is to farm, even when the weather cooperates. In addition to keeping the vehicles and implements up and running,

there's the outlay for herbicides, seed, fertilizer, and so forth. The first wool clip from our sheep covered the cost of chemical drenches and shearing, but it was clear the earnings weren't going to be nearly enough to support the growing community. What's more, the farm was placing unsustainable demands on our workforce at a time when we were also putting up buildings and finding ways to get involved in our new neighborhood.

So we started a sign-making business to provide an income. That's developed well, so that we're now able to support the two hundred twenty people living together in community here.

Why not just walk away from farming?

The Bruderhof's way of life has always been linked to the land, going back to early pioneers like Philip Britts (see page 58). For our first four decades of communal living, farming literally put food on the table; since then, we've earned our money mostly from manufacturing. So the establishment of the Danthonia Bruderhof twenty years ago was a chance to connect again with the land. Over time, with adjacent acquisitions, we brought our farming country here to around 5500 acres, of which perhaps half was pasture, a quarter cultivated, and the rest wooded slopes and rough country only marginally suited for grazing.

Still, we weren't coping. So we scaled back and tried a few common models. We allowed contractors to run their cattle on our property for a fee. For a number of years we leased out the farm. Of course, as we discovered, tenants naturally want to make as much money as possible from the property, often at the expense of the landscape. Within a couple of years, overgrazing had seriously damaged our land.

My family arrived at Danthonia in November 2004. By then the effects of

All photographs courtesy of Danthonia Bruderhof

what would become known as Australia's Millennium Drought were already painfully obvious. Drought became even more severe, persisting until autumn 2010. The drought caused us to think hard about the way we cared for our landscape. I'd come from England, where rain is often more of a bother than a blessing. So it was extremely strange to find myself constantly looking to the west and watching the clouds every day, month after month, year after year, waiting for the gift of rain. When occasional rain did fall, the land hardly responded – it was in such poor condition.

In 2007, we watched our creek dry up. It's a beautiful little creek, lined with willows, running through a wide floodplain at the base of Swan Peak, Danthonia's prominent landmark. I will never forget seeing the drying, algae-glutted pools, with dead fish – golden perch, eel-tailed catfish, and Murray cod, some up to thirty inches long – lying belly-up; the eroded, crumbling banks; no flow whatsoever. In 2009, it happened again, only with no dead fish. They were all gone. I remember thinking, "This isn't right. What are we doing here?"

So the crucible of drought turned us to an arduous process of discovery: how to bring our landscape back from the brink and restore it to life and health.

Yet was the Millennium Drought really such an exceptional event? Isn't drought just part of how Australia's climate has always been?

That's only partly true. There's no doubt it's a rugged environment – you can plan on droughts and floods. Farmers in our region will tell you that for any ten years you can figure on two bumper years, three or four reasonable-to-mediocre years, and three or four total failures.

But now, in addition to Australia's naturally fickle weather, we're also contending with climate change: average temperatures have risen markedly since 1950, with an even greater increase in how often we get days with extreme temperatures. That certainly increases the challenge of farming.

It's crucial to remember we're part of a much longer story. Australia is home to the world's most ancient continuous civilization, with the first indigenous Australians thought to have arrived around sixty-five thousand years ago. If you equate that span of time to a single day, then the first European settlers – the First Fleet that entered Botany Bay from England in 1788 – got here less than seven minutes ago.

Some of those Europeans kept excellent diaries or made drawings and paintings of

Aerial views of Danthonia in 2000 *(left)* and 2017 *(right)* showing the community's growth in a healing landscape.

Paddock and dam before and after holisic management: in 2007 *(left)* and 2015 *(right)*

what they found. The country they describe was vibrant and healthy, with natural pastures boasting three to four hundred species of plants. They documented that even when no rain had fallen for three months, the valley systems held lush, fresh grass.

Their records clearly show that the fertile zones were not limited to the coastline. They describe a landscape beautifully adapted to benefit from the cyclical climate, with unique functions that enabled it to capture and store water. They described topsoil depths of one to two meters, with cracks deep enough a man could hold a machete and lower it and his entire arm into them, and still not touch the bottom. That soil was so spongy that even during a drought a single wagon's tracks would remain visible in the pasture sward for years.

The explorers found a landscape that functions very differently than Europe's, whose large waterways carry superfluous water to the sea. Instead of these, Australia had wide, broad floodplains filled with reeds over a dozen feet high, swamps with occasional pools, and riffles where water moved downhill through the valleys. Some of these riparian systems were twenty-five miles wide and capable of holding immense quantities of water, which was released into the landscape during drought. Others, such as the marshlands along Danthonia's creek, were narrower, but served that same

sponge-like function.

It's hard to overstress the importance of the plants in these systems. The reeds and other diverse, multi-storied plants managed water for the landscape. Tragically, that was not understood by the first Europeans. They brought their own paradigms: drain the swamps, open up the waterways to boats, bring hard-hooved grazing animals into the valleys, plow the topsoil and plant monoculture crops.

The results were catastrophic. In less than ten generations, Australia has seen massive erosion and desertification brought on by the destruction of functioning riparian areas and by farming practices that disregard the landscape's natural ability to hold water and keep salts at bay. Today, we're farming on subsoil, not topsoil. Natural plant and animal diversity is a shadow of what it was. With few plants to help store water in the landscape, slow its movement, spread fertility across flood plains, and control salts, when rainfall comes it washes out to sea, carrying with it untold tons of precious topsoil.

Yet some would say you're painting far too bleak a picture. Is Australian agriculture really in crisis?

I can only speak from the experience of the farmers here in eastern Australia. As has been widely reported in the media, farmers are really struggling – as a result of the drought, they have little remaining ground cover, livestock are starving, wildfires have increased, and there's dust blowing everywhere. Many farmers say it's the worst drought in a hundred years. They've had to drastically reduce the number of animals they stock and have been hand-feeding hay and grain for the last one to two years.

This situation spells financial ruin for many. The fact that our government agencies have been giving out hundreds of millions of dollars in emergency aid to farmers indicates the scale of the crisis. And for the individual farmer, the aid offered is often pitifully inadequate.

From what I've come to understand from numerous farmers across this country, as well as from the insights of scientists and farmers internationally, there is no doubt in my mind that this is largely the result of conventional farming methods. These have done great damage, especially here in Australia.

Still, haven't the methods of agriculture developed over the past hundred years been hugely effective in growing more food?

It's true that through mechanization and artificial fertilizers, we massively increased our ability to produce food. On the face of it, this looks like amazing progress – much more food for a lot less labor. But what becomes ever clearer is that industrial agriculture violates the natural ecosystems on which human beings rely. It hurts the landscape, the animals, and us, the consumers. And as we're seeing here in Australia, at some point it stops working.

Take intensive tillage, for example. Plowing, disking, and harrowing expose soil biology, a vital source of plant nutrients, to the elements. Moisture evaporates, soil erodes, and soil

carbon is lost. In our part of the world, our soils are vertisols – heavy, cracking, clay-based soils that are very prone to erosion. In one good thunderstorm, inches of topsoil disappear from tilled ground, as we saw at Danthonia in our early days of Australian farming.

With industrialization came monocultures – huge fields planted with a single crop variety. Every farmer implements crop rotation, but when we ignore the contribution diverse plants make to soil biology through root exudates (more on those later) and load on synthetic fertilizers, which can eliminate root exudates altogether, we're killing our soil.

After World War I, huge factories that had been cranking out ammonium nitrate for explosives were suddenly shuttered – but not for long, because someone discovered that ammonium nitrate was a wonderful source of nitrogen for plants. Soon they were synthesizing the three primary nutrients for plant growth – nitrogen, phosphorus, and potassium – and ladling this NPK fertilizer onto crops.

And the plants seemed to love it. They grew higher and yielded larger harvests. But the soil suffered. In a healthy ecosystem, plants live in symbiosis with the soil in an exchange of sugars for nutrients. What we didn't realize was that by short-circuiting the nutrient cycle through NPK fertilizer, we were cutting off numerous other nutrients and micronutrients that would come into the plant and build a nutrient-dense product. As a result, we now have woefully nutrient-deficient crops. In fact, studies by the research scientist Donald R. Davis suggest we may need to eat far more fruit and vegetables than our grandparents did in order to get the same nutritional benefits. There's strong evidence linking these dietary deficiencies to our increased health problems.

On top of that is the witch's brew of chemicals we've dumped on our landscape for years,

Restoring a Creek

JOHANNES MEIER

1 **2007:** We began with a heavily eroded creek bed with only minimal plant life. Trampling of the shoreline by conventionally grazed cattle worsened erosion. During flood events, swiftly moving water caused further erosion and carried off topsoil and nutrients, while lowering the water table still more.

2 **2015:** The same location as in the 2007 picture *(left)*. Now reeds and other plants in swampy areas slow down the water while helping suppress salinization and providing a habitat for animal life.

3 **Contour banks** feed flood waters onto flood plains through braided channels. These flood plains function as a reservoir of water in the landscape.

4 **Leaky weirs** – porous obstructions made of natural materials – slow down first-order streams. They serve to de-energize the water current and allow the water to seep into the soil.

5 **Fencing** keeps cattle from eroding creek banks and damaging the creek bed and the plants holding it all together.

6 **2016:** Working with nature, we have seen the creek rebound and begin to show hints of what it must once have been like before European settlement: a biodiverse and scenic source of fertility for the surrounding landscape.

with limited understanding of the toxic effects they have on our ecosystem and on us. Our Western diet has become a delivery system for poisonous chemicals. Glyphosate (the key ingredient in the weed killer Roundup), which is increasingly suspected as a carcinogen, is everywhere from children's breakfast cereals to German beer, some of which has been found to contain three hundred times the legal limit for drinking water. A 2016 study concluded that 93 percent of Americans have glyphosate in their urine. To what extent are agricultural chemicals contributing to the increase in a wide range of diseases in Western countries – autoimmune disorders, obesity, diabetes, heart disease, infertility, and autism?

Intensive irrigation is another aspect of industrial agriculture that has often proved destructive, especially here. It's opened farming opportunities in landscapes with inadequate rainfall, yet these areas often have high levels of salt in their groundwater. Over time, through irrigation, salt build-up in the soil has ruined huge swathes of agricultural land. Aquifers can't keep up with our insatiable need for water, so perennial drought ensues. In Australia, agriculture accounts for 50–70 percent of the nation's water usage but only 3 percent of GDP. Recently, Australian media have extensively documented the ecological disaster that is the Murray Darling River system, brought on largely by mismanagement of water for agriculture.

The good news, however, is that there is a way back. We can stop destroying and start restoring. We can work with nature so that, in essence, the land heals itself. It is simply a matter of grasping certain principles that must be respected. For us, coming to terms with these key principles was a journey. But we now know there are logical steps to follow.

Walk us through how that actually works. Danthonia has similar rainfall to its neighbors, and yet here there's feed for the cattle and water in the dams, while neighboring farms are parched and the livestock is starving. Why the marked difference?

There's a multifaceted answer. Let's start by looking at what makes a landscape healthy.

One indicator of an ecosystem's health is the ability of its plants to convert sunlight to sugars. These sugars help plants grow and are also fed to the soil, transforming into humus – that amazing substance at the heart of topsoil that holds minerals and resources, as well as four times its own weight in water. Through the plants' activity, humus and soil organic matter are built up, increasing soil carbon, and the land begins to absorb and hold more moisture. Instead of running off the hillsides to the ocean, rain stays in the landscape.

Scientists calculate that for every percent of soil carbon present, the landscape can hold 140,000 liters of water per hectare (about 15,000 gallons per acre). In Australia, the estimated average soil carbon content pre-European settlement ran from 7–20 percent. It now averages below 1 percent. Imagine if we increased it to 5 percent, which is still below pre-settlement levels: ten acres of land would be capable of holding enough water to overflow an Olympic-size swimming pool.

Understanding this is one thing. But how do we actually bring the landscape back to health? Here at Danthonia, having seen how overgrazing by tenants had damaged our pastures, we knew the answer would involve the way we run cattle. Through that, we learned about the concept of holistic grazing management and its proponent, Allan Savory. He is a Zimbabwean ecologist,

Planting olive trees with young and old

environmentalist, and livestock farmer who set out to discover why so-called overgrazing was ravaging the African landscape when herd sizes were at an all-time low. Eventually he struck on the fact that hunting had decimated the predators that historically forced herds to stay tightly bunched for protection. His experiments demonstrated that when cattle were bunched together to mimic traditional herd density, the landscape began to revive.

The wildebeest on the savannahs of Africa or the buffalo on America's Great Plains did more than just consume and trample grass as they moved in large, tightly bunched herds, harassed around the edges by lions or wolves. Rather, by thoroughly speed-grazing an area and enriching it with their manure before moving on, they provided plants the means to grow to full potential. And we know that those areas were fertile; in the United States, meters of

topsoil were built up this way, only to be tilled up and blown away during the Dust Bowl years.

That's holistic grazing management in a nutshell. Once we understood its basic concepts, we got out our fencing tools and began dividing our large paddocks into smaller ones, where we crowd more cattle for shorter grazing periods. Holistic grazing requires careful planning and recordkeeping and the commitment to move cattle frequently, sometimes daily. In my opinion, it is our most powerful tool to exert landscape change and recovery. That is what regenerative agriculture, as we call it, is all about.

How does that help in combatting the effects of drought?

Around 2006, we came across an Aussie named Peter Andrews, who's given forty years of his life to understanding how to regenerate the landscape. Peter is a genius for his ability to read landscapes and understand the functions needed to restore them. He developed his ideas into a framework called Natural Sequence Farming.

I visited Peter's farm in 2007, at the height of the Millennium Drought. He took me first to the neighboring property and showed me the bone-dry creek. Then we walked downstream onto his land. Pretty soon there were pools of water, and floodplains growing green. By the time we reached the end of his property, the creek was a flowing stream. It was astounding: in the middle of a desert-dry landscape was a running creek surrounded by dense trees and shrubs, with abundant animal life. We continued along the creek into the adjoining property. Hardly three hundred meters downstream from Peter's farm, the bed was dry again. I had never seen such a stark demonstration of the link between the local ecosystem and water in the landscape. It inspired us at Danthonia: if Peter Andrews can do it, why can't we?

With Peter's input, we started implementing Natural Sequence Farming. We removed our cattle from riparian zones to allow the creek to heal and to encourage vegetation along its banks and floodplain. In these areas, we planted numerous trees. In times of flood, this plant life helps slow the water and capture nutrients. The goal is to encourage the natural recreation of linked ponds and reed beds, such as were here for millennia.

On the slopes higher up we are working to mimic other once-natural functions, by building level contour banks that retain water from big rains. When the bank overflows, strategic openings allow water to spread slowly so the land can absorb it. Nutrients washed down the slope also spread evenly. Below the bank, we plant trees that utilize those nutrients, provide shade, and contribute to soil biology. In our valley, which drains to the creek, we're developing a system of ponds and reed beds and planting trees to slow the movement of water.

The community has planted thousands of trees over the past two decades. Why?

So far, we've planted around one hundred thousand trees. They bring all kinds of benefits. Trees impede wind as it travels across the landscape; the faster wind moves, the more moisture we lose. Trees provide habitats and shade. Where there are trees, the earth will absorb up to sixty times more rainfall than pastureland. Their roots pull nutrients up from far below the surface – a mature tree deposits 7 percent of its full biomass into the soil every year, which benefits shallower plants. And they are simply beautiful.

We've planted trees strategically, often in beltways along ridges. These are a mixture of native and exotic varieties for diversity. We fence them to keep the cattle out. Today, trees we planted fifteen years ago are tall enough that cattle and native wildlife can move among them, enjoying their shade and depositing their dung and urine high in the landscape where the nutrients can do the most good as water moves them downhill.

Have you seen measurable changes yet as a result of the steps you've taken?

Definitely! Birds are an early indicator of ecosystem health. When Danthonia's bird-watchers started counting over a decade ago, we recorded around one hundred bird species; today that number stands at 150 and counting.

Photograph by benjamint444

Red-capped robin: since 2006 birdwatchers have seen five to six new species each year, even in dry years.

You spoke earlier about the incredible loss of topsoil from Australia's landscape. That soil must have been thousands of years in the making. How can you hope to regenerate it?

Of the fifty new arrivals, eleven species are associated with expanded and healthier wetlands and open water, and fifteen with the significant increase in flowering trees and shrubs. The rest relate to species migration and generally improved quality of life.

Three months ago, as a result of the drought, the contour banks dried out. Yet the sward is spongy underfoot, and the grass is green and continues to grow. This underscores that the most effective place to hold water is in the soil. Recently I compared our well water records against our current usage. Despite very poor rainfall, our wells have higher water levels than in bygone droughts. More water is being held in the landscape.

At our creek we've found that water enters our property at a flow rate of sixteen liters a minute. Where the creek leaves our farm, water flows at fifty liters a minute. Even during drought, the volume of water we're passing to our neighbors downstream is three times greater than what we're starting with! It's because water is being retained higher up on our land and working downhill over time. This is tremendously encouraging and inspires us to keep working.

Historically, some experts have said it takes three hundred to a thousand years to build inches of topsoil. But we've learned that's actually not the case. Soil is not built primarily by decaying leaf matter and so forth. Living, healthy topsoil is created by plant root exudates – the carbohydrates, vitamins, organic acids, and other nutrients released into the soil by the root systems of plants. Of the sugars that plants create through photosynthesis, 30–40 percent transfers to the soil through the roots in exchange for nutrients. In this way, plants feed soil biology: fungi, bacteria, microorganisms, and mycorrhizae, the symbiotic associations between plants and fungi in the root zone. Those take the sugar and convert it to humus, which is topsoil.

So topsoil can actually be built quite quickly. But it won't happen without diverse plant life. This diversity is key, and it has everything to do with the way we farm.

It's a cutting-edge area of scientific research. We're learning that as plant diversity increases, there's a certain trigger point – called quorum sensing – where topsoil begins to build rapidly. How many species of plants do you need for a quorum? The more the merrier, microbiologists are saying. Different plants produce different root exudates, allowing access to specific soil nutrients. There have been positive results from as few as twelve species, and more rapid success with forty.

Our best native pastures at Danthonia contain fifteen to twenty species. We're a long way from the hundreds of species these landscapes once enjoyed – all forming a richly diverse pasture sward that allowed topsoil to build and be maintained, enabling it to hold water and release it during drought. The challenge is that it's tough to grow a diverse pasture on poor soil. Spreading compost extract introduces living biology into the soil, but that biology then struggles to survive. So we're experimenting with a biological stimulant, a cocktail of microbes and organic compounds that feeds and encourages the living elements already in the soil.

One of the reasons we are determined to improve plant life across our property is to combat salinization. In Australia, every rainfall adds salt to the continent: salt isn't washed out to sea because of the absence of large river systems, so it accumulates, and unless suppressed, it will kill the soil. In bygone times, swampy reed beds and a range of other plants held water in the landscape. This created a subsurface freshwater lens that suppressed the salt and kept it from rising and damaging soil biology. We're working our way back to that, but we need healthy soil for that to happen.

Where does this leave farmers who rely on crops for their livelihood?

I can't speak from personal experience, since at Danthonia we got out of cropping a decade ago and converted the land into pasture. But there are promising models of a different way to farm crops. In Western Australia, for example, Ian and Di Haggerty run their farm without chemical inputs. They graze livestock intensively, then sow wheat into those same paddocks and grow it over winter for spring

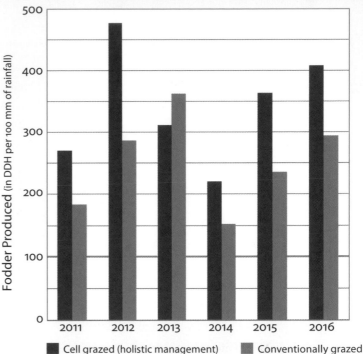

Cell grazed (holistic management) ■   Conventionally grazed ■

Pasture Productivity

This chart shows the average productivity of four test sites on the Danthonia farm over six years. Each site was divided into two: half was grazed using holistic management, with cattle being moved between smaller paddocks every 1–5 days, while the other half was grazed conventionally. Cell grazing allows plants more time to recover, strengthening root systems, improving soil health, and increasing yield.

harvesting. Once they've harvested the crop, they move the cattle back in. Working with the livestock, they're able to control weeds without chemical sprays. Their crop yield is less than if they used synthetic fertilizers, but it also costs less to grow, so they're coming out ahead. At the same time they're building soil carbon and biology.

What do you say to conventional farmers who are open to changing but may feel it's too costly to transition?

Take one step at a time. If you're into livestock, take steps toward management-intensive grazing. If you're into cropping, seek out successful regenerative farmers and learn from them. I tell them, too, that the first few years transitioning may be tough, but the results speak for themselves. Cost inputs will go down, nature will begin to heal. The good Lord has designed a natural system that is remarkably able to recover from the worst we've thrown at it.

What would you like to see Australia's government do in support of farmers?

Number one, there has to be buy-in for a regenerative approach, as a matter of urgency. We need to educate officials to convince them to move beyond time-honored but harmful paradigms. Number two, we need government funding to help pay for farmer education and for the upfront costs involved in transitioning to regenerative agriculture: for instance, the fencing required for rotational grazing.

Perhaps the biggest point of contention is water use and the way it's regulated. Current laws are designed to facilitate moving as much water as possible off the land and into storage areas, from which it can be used for irrigation. That's counterproductive and needs to change. All of this has to be done responsibly; teamwork among scientists, politicians, bureaucrats, and farmers is essential.

Have you run into opposition to regenerative agriculture?

Sure. Recently I went into our local store and was talking to the agronomist there, whom I've known and got along with for years. I said, "I'm no longer going to be using herbicides." He just turned and walked away. That was the end of the conversation. Whether it's about money, I can't say, but he makes his living selling herbicides.

Tim Wright, a farmer down the road from us, has spent twenty-five years regenerating his property. He's been very successful. Yet despite that, he has neighbors who still ignore him, and he's the butt of jokes at his local store. We farmers can be a hardheaded bunch – but I like to think that can work in our favor when we put our minds to fixing a problem!

Over time, the Danthonia Bruderhof has earned the trust of the local Aboriginal community. Is there a connection to how you farm?

In Australia, there's a tradition of beginning public gatherings with a "Welcome to Country": an indigenous person acknowledges the "traditional custodians" of the land on which the gathering takes place and pays respect to elders past and present. This concept of custodianship – of being caretakers of a landscape, so as to pass it on to future generations in better shape than we inherited it – is one we have tried to embrace. So that is common ground, I believe, with our Aboriginal neighbors.

The truth is, what we're doing at Danthonia to care for the land is not such a big deal. As thrilled as I am with the steps we're taking and the way nature is responding, it is only one small part of why we live in community. Our calling is to live a life of discipleship of Christ, and to follow his path as best we are able. Caring for this land is simply a reflection of our desire to be true to Christ who loves the flowers of the field, sparrows, children; who takes pity on the sick and needy; whose heart is with the destitute and downtrodden. This impels our efforts toward fellowship with and understanding of those around us, and particularly with our Aboriginal brothers and sisters. Charles Massy, whose book *Call of the Reed Warbler* is an absolute must-read for anyone considering regenerative agriculture, says it well: "Not until we attain reconciliation

Johannes addressing 300 local farmers gathered at Danthonia in September 2018 for a field day on regenerative agriculture. Also participating was soils ecologist Dr. Christine Jones, whose research into restoring topsoil with plant life has inspired Danthonia's work over the last decade.

with both the land's first peoples and the land itself, will we be enabled to 'arrive' and truly belong on this continent."

Regenerative agriculture is ultimately about getting back to the task that God gave Adam and Eve in Genesis 2: to care for the earth that God created as his garden. We have to start humbly, recognizing that we Western consumers are complicit in the global ecological disaster that industrial agriculture has created. Greed and demand drive the markets – and separate us from the way God intended us to live. Agriculture has a lot to answer for, but I truly believe that agriculture has tremendous potential to regenerate our world.

Those of us who farm need empathy for the land and animals; for our neighbor next door;

for those starving a world away; for future generations; for everything God created. Our hearts must work as hard as our heads and our hands.

Recently I was reminded of words by the prophet Jeremiah, who lived in desperate times. "This is what the Lord says: 'Stand at the crossroads and look; ask for the ancient paths, ask where the good way is, and walk in it, and you will find rest for your souls'" (Jer. 6:16).

All that's required of us is to recognize the crossroads, ask to see the ancient and good way, and then step in that direction. ⤳

Interview conducted on January 11 and 17, 2019, at Danthonia Bruderhof in Elsmore, New South Wales, Australia.

Love Is Work

EBERHARD ARNOLD

Sybil Andrews,
Haulers,
1929

WORKING TOGETHER with others is the best way to test our faith, to find out whether or not we are ready to live a life of Christian fellowship. Work is the crucial test of faith because such a life can come into being only where people work for love. Love demands action, and the only really valid action is work. Christian fellowship means fellowship in work.

JUSTICE AND LOVE DEMAND that everyone take part in simple practical work with a spade, hatchet, or rake. Everyone should be ready to spend a few hours each day in either the garden or on the field: digging and spreading manure, plowing, or hoeing potatoes; on the reaper, at the circular saw, or in the metal-working shop. Everyone should be ready to devote a few hours every day to this practical work; those who have done purely mental work till now will feel its humanizing effect especially. In this way it will be possible for each person's unique gifts to be kindled. The light that flickers within each heart will then exhibit its once-hidden glow in scholarly research or in music, in expressive words, in wood, or in stone. ➤

Source: *Eberhard Arnold: Writings Selected with an Introduction by Johann Christoph Arnold* (Orbis, 2005).

Eberhard Arnold (1883–1935), a German writer and theologian, was Plough's *founding editor and a founder of the Bruderhof communities.*

Letter from North Carolina

RICHARD JOYNER

Gardens teach freedom. Here in Conetoe, North Carolina, our community is surrounded by a food desert on all sides. A quarter of households in our county live below the federal poverty level. Poor nutrition has caused more young deaths here than car accidents. I've been a hospital chaplain for years, accompanying one family after another as they stand at the graveside of someone who died too young. I remember the day on the way home from the hospital that I pulled my car over and prayed. It was like I heard God saying, "Open your eyes and look around." There were fields flowing out in every direction.

But there's been a hard past in those fields. My parents and grandparents were sharecroppers. The generation before them were slaves. Those fields didn't spell freedom for me. But now a new vision began to overwrite the past. What if our community could turn this around, find our freedom in good soil, and start to believe in a future where families can survive and thrive through determination, creativity, and last year's seeds?

Conetoe Family Life Center has been running for fourteen years now. Young people in our program build social and spiritual skills as they discover what it's like to be one with the land. Yes, we learn how to grow everything that can grow in this soil. But we're also learning how to invest in the community, putting food on family tables as well as selling vegetables, fruit, and honey to local restaurants, hospitals, and school systems. Then we reinvest those proceeds in the ground.

I want the children to know, "I can do this! I can contribute to my family's and community's well-being." We've gone from being a dangerous food desert to being a community that can feed hundreds of families each week without relying on outside sources. Our financial income levels have not changed. But what we do with our shared strength means that our net income has changed. We're not dependent – we can speak and work justice. If a child knows that because of work and community, the lights aren't going out, the food will be on the table, the roof will stay over his head, that's justice. The early mortality rate is dropping steeply, and that's justice.

When our community gathers for a festival, we're celebrating our own roots. We acknowledge pain and injustice; we know there is much that still needs to change. But we also give thanks for the chance to grow God-given food, and to eat it in peace together. Here in Conetoe, gathering around a table is a life-saving action. ≥

The author at Conetoe Family Life Center with coworkers

Rev. Richard Joyner is pastor of the Conetoe Chapel Missionary Baptist Church in North Carolina, where he also runs the church's community garden. He was named a 2015 CNN Hero.

PHILIP BRITTS

How Shall We Farm?

In this far-sighted essay from the 1940s, a Devonshire-bred farmer sketches out a sustainable way of agriculture – and of life.

Winslow
Homer,
*Man with
Plow Horse,*
1879

WITHIN THE LAST half century we have seen a revolution in agriculture, typified especially in the rising use of chemical fertilizers and powered machinery. The change has brought consequences both good and bad. The tractor and the fertilizer bag are really means of speeding up farming – the first of speeding up man's powers of cultivation, the second of speeding up nature's supply of nutrients to the

crops. They were called into being through the increased pressure of machine-age life upon the land, to supply the ever-increasing demand for more food, more oils, more fibers, with dwindling manpower. This demand has been largely met, and could have been more nearly met but for the ruinous periods of destruction in two world wars. But a price is to pay. Where this speeding-up has been done rashly, as it has on millions of acres of land, retribution has

---

Philip Britts (1917–1949) was a horticulturalist and poet. This undated essay is excerpted from Plough's *book featuring his essays and poetry,* Water at the Roots: Poems and Insights of a Visionary Farmer *(2018)*. plough.com/britts

followed swiftly, and good farming land that nature took millions of years to form has been worn out and lost in a generation.

All farming, even in its most primitive form, is necessarily an interference by man with the process of nature. As agriculture has developed, this disturbance of nature, this effort at control over nature, has increased. Adam was charged with the double task to "subdue and replenish" the earth. If a graph could be plotted of the subjection of nature by man, it would show a line, rising slowly at first, through several thousand years, then abruptly and very steeply in the last few years. A graph of the replenishment of the earth by man would probably show a slow rise throughout the centuries, but instead of following the sharp rise of the line of subjection in modern times, would perhaps curve downwards. This in spite of the extensive use of fertilizers, because chemicals without humus do not give lasting or balanced replenishment.

Where will the lines go from now on? Obviously if the measure of subjection continues to rise, and the measure of replenishment falls, if the lines get farther apart, nature will rebel, and bring down the measure of subjection by such hard steps as erosion, sterility, and disease.

Here and there are signs that some people are awake to the trend of things, and alarmed at the possibilities. Notable are such men as Howard with his "Indore process" of composting, Pfeiffer with his "biodynamics," Faulkner with his *Plowman's Folly.*[1] A drive for sounder farming practice is represented by a section of the British landowner class, who have been vocal in the House of Lords, and from time to time evidence is submitted by scientists and technicians from Experiment Stations.

The cause of good farming is soundly served by such careful studies as *Humus* by Waksman, and *Soil and the Microbe* by Waksman and Starkey. But these books, thorough and sound as they are, are highly technical. They leave a gap to be bridged between the scientist and the farmer.

A still greater gap exists between what the farmer knows to be right, and what a competitive economy forces him to do.

Experiment Stations supply the farmer with valuable fractions of knowledge, which he has to fit into his scheme of farming. The results of experiments, and the elaborate methods for determining the "statistical significance" of the results, are not above question. However carefully they are planned and executed, however often repeated, they cannot give an infallible verdict on the effect of a practice in the long run, or even an infallible prophecy of the effect of a practice in the next repetition of it under slightly different conditions. At best they can only give evidence that certain effects do recur within a finite margin of variation.

But if one makes claims for, or passes judgment on, technical matters without the *restraint* of these "cut and dried statistics," several human tendencies must be guarded against. One is the tendency to overgeneralize, leading either to a one-sided championing of one direction or the other, or, equally false, to the compromise of taking a "middle way," whereas the technical truth for each specific case may follow a line swinging from side to side within a channel formed by a series of correlations. In farming, as in life, isolated factors lose their significance – it is the correlation of things that matters. Another expression of overgeneralizing is to "give a dog a bad name and hang him" – for example, to attribute the use of fertilizers to an "NPK mentality" and dismiss the subject.[2] . . .

One must beware of attributing to any one factor effects caused by a number of factors. Thus soil fertility is *not* the basis of public

> A tragedy of the modern world is the divorce of city dwellers from the land.

health. It is one of many factors and, I think, a major one. But food by itself, however good, will not produce health. We have to cope with other factors such as climate, hygiene, and moral and social habits. Even if all the factors we can discover were made to contribute towards health, it would be fallacious to suppose a disease-free mankind. There are deeper causes of disease and death.

In the same way, organic farming by itself, as an isolated factor, can never reach its full significance. It stands for the great truth that agriculture can never be stable and permanent until man learns *and* obeys the laws of fertility, the cycle that includes the decay of the old and the release of the new, or, to put it biblically, to "subdue and replenish."

But it can only realize its full meaning in the context of an organic life. Man's relationship to the land must be true and just, but this is only possible when his relationship to his fellow man is true and just and organic. This includes the relationship of all the activities of man, the relationship of industry with agriculture, of science with art, the relationship between the sexes, and above all the relationship between man's spiritual life and his material life.

One of the great tragedies of the modern world is the complete divorce of the city dwellers from nature and the land. Civilization has become like the "fat white woman" who "walked through the fields in gloves."[3]

The decisive factor in the success of the farmer will be, ultimately, the love of farming. This love comes when we find, not in nature, but through and behind nature, that something which impels worship and service. Part of the glory of farming is that indescribable sensation that comes, perhaps rarely, when one walks through a field of alfalfa in the morning sun, when one smells earth after rain, or when one watches the ripples on a field of wheat. . . .

This love is fed by understanding, by knowledge. Without going the whole way with Leonardo da Vinci and his "perfect knowledge is perfect love," the more one knows of the mysteries of the earth the better one can love farming in the sense of giving one's service to it.

One of the best contributions of the school of organic farming to agriculture is this call for a genuine love of the land.

Our communities in England and Paraguay represent an effort to extend this love of the land, of organic farming, to all the other aspects of life: to industry, craftwork, education, and the daily relationship of man with man.

Sociologically, the position of our community in Paraguay is an interesting one. We find ourselves a group of modern people transported to a land that is still only at the initial stage of the machine age (i.e., before the sharp rise in our supposed graph of the subjection of nature). We find ourselves setting out to farm with a modern scientific outlook, but without modern scientific appliances. What implements we have are of the pre-machine-age type, pushed by hand or drawn by horses.

We are faced with the necessity of feeding ten people from every eight acres that we are able to cultivate or stock, including maize, beans, *mandioca* (the potato of Paraguay), peanuts, fruit, vegetables, milk, and eggs. We are unable to imitate the Chinese with their unlimited expenditure of labor to maintain fertility, because we are always short of manpower. For this reason meat plays an important part in the diet, being cheaply produced on the native range.

To maintain the fertility of our land we use green-manures, for which we have plenty of scope in our nine-month growing season, and two-year leys grazed by the dairy herd, combined in a five-field rotation. During the short winter, those fields that are empty are protected by leaving the summer cover crop on the surface as a mulch, or by disking in rye, to be lightly grazed and plowed under for green-manure.

We do not use fertilizers because they are too expensive, but believe we could advantageously use bone meal and cottonseed meal if they were economically obtainable. In the same way we would welcome the chance to use tractors if we had the money to buy them, which we have not. We do not, with Faulkner, reject the use of the plow. Although we believe there is such a thing as over-plowing, we still believe in "the useful plow," providing one never plows without turning under organic matter, and never leaves the plowed land naked to the weather in this climate of heavy rain.

We have only been tilling the land here for five years, but with the number of people supported per acre, and two crops per year, the demands on it have been heavy and are increasing. We are therefore vitally interested in the maintenance of fertility.

But just as we do not believe that organic farming can find its full meaning outside the context of the whole of life, neither do we believe that an organic society can exist for itself, or have its only significance for the small group of people who are living it.

One of two things must happen. Either man will decline, through war, famine, disease,

> # We do try to farm organically, but we see this as only a part of an organic life.

and the falling birth rate – and the recent progress of science leads one to believe that this decline may be imminent and rapid, and accompanied by obvious horrors – or we must learn to live peaceably together, in a society where the demand for wealth or position, ease or comfort, is supplanted by the just sharing of everything, and a free giving of strength and brains in service, not of self, but of the whole.

We do try to farm organically, but we see this as only a part of an organic life, and existing in the context of a search for truth along the whole line. This gives rise to social justice as brotherhood, to economic justice as community of goods. We see these conditions as the necessary basis for a true attitude towards the land and towards work. Therefore our door is always open to all people who wish to seek a new way with us.

To the question, "How shall we farm?" must be added the question, "How shall we live?" ⟶

1. Albert Howard (1873–1947), Ehrenfried Pfeiffer (1899–1961), and Edward H. Faulkner (1886–1964) were pioneers in organic farming methods.

2. Howard's work is the story of "a Fall" in which the "serpent" is Baron Justus von Liebig, who showed that plants need only nitrogen, phosphorous, and potassium to grow. Howard called this the "NPK mentality" after the chemical symbols for those three elements.

3. The reference is to a poem by Frances Cornford: "To a Fat Lady Seen from the Train."

AIDAN HARTLEY

Cows and Elephants

Conservationist ranching in the African savannah

WHEN WE MOVED to northern Kenya to develop a modern farm early this century, what I first thought I saw was a pristine Garden of Eden. On high plains I walked among an abundance of wildlife – predators and herds of elephant, zebra, eland, oryx, giraffe, and antelope. Here were savannahs and open woodlands, an arid wilderness without fences.

At first I imagined that any type of human activity I embarked on as a farmer would ruin this paradise. But I was ignorant. Our land was an Edenic wilderness only at first glance. On the horizon, my neighbors were pastoralists, tribespeople who tended cattle, sheep, goats, and camels in a semi-nomadic lifestyle going back millennia. Or they were modern commercial cattle ranchers, most – like me – descendants of colonial British settlers. Or they were conservation groups such as The Nature Conservancy, which invests in this part of Kenya to establish sustainable ways of protecting wildlife and providing impoverished communities with education, health, and job opportunities.

I discovered human marks already etched across the landscape. Prehistoric hand axes and obsidian tools littered the ground. Out rambling, I stumbled on cairns – stone circles, burial mounds, monuments – some isolated, others in sentinel rows. They overlooked views of mountains or water, to me a sign across millennia that we share a love of nature. On local cave walls were tableaux of wildlife, human figures, and cattle. For these were some of the first pastoralists.

From about 3000 BC in the district, our hunter-gatherer ancestors began tending cattle, sheep, goats, and camels. The landscape was forested until then, but livestock keepers used fire to create open grasslands. Livestock fertilized soil, improving pastures that now sustained not just domestic animals, but also multiplied the wild mammals we associate with today's African savannahs.

Almost until the present day, pastoralists and livestock shaped our local landscape alongside wildlife. Only with the arrival of colonial British ranchers did that change. Farmers regarded wildlife as competition for grazing. During the 1940s wild creatures were wiped out to feed Italian prisoners of war in camps around the base of Mount Kenya. A decade later, writer Elspeth Huxley recorded that all lions in our area had been exterminated.

Attitudes reformed in the 1980s. Cattle ranching in Africa is tough, and farmers saw the advantages of conserving wildlife because they could start tourism businesses to supplement meager incomes. Conservation was becoming broadly popular in Kenya, where sport hunting had been banned. If you leave nature alone, it tends to recover swiftly – and now that people were not out to relentlessly kill them, lions and other creatures naturally recolonized our landscape. Within a few years, ranchers and pastoralists were living in relative peace alongside predators and other game. In fact, our commercial ranchers and relatively poor cattle-keeping tribespeople are more tolerant of lions than farmers in North America are toward wolves being reintroduced in national parks.

During the days our livestock are herded across the open grasslands, since we have no paddocks. At evening they are brought into a *boma*, or night enclosure, protected by a stockade of acacia thorns and drystone walls. We have learned to adopt traditional pastoralist methods, and if left out in the field overnight I would say my cattle have a 50 percent chance of being killed

Our land was an Edenic wilderness only at first glance.

and eaten by predators. As it is, in fifteen years of keeping cattle, I have lost only one calf to a lion during the day, while a handful have been killed at night. These are losses I can tolerate – though I found it harder to take the slaughter of twenty-five sheep in leopard raids during a string of nights over the last year. While breathing deeply, I try to compare this disaster to having a fox in a chicken coop in Europe. That does not cheer me up entirely, because our poultry are already harassed – not by foxes, but by the occasional giant spitting cobra: eight feet long, highly venomous, and hungry for eggs.

Our sometimes predator-harassed farm is semi-arid, at high altitude, with cyclical

Aidan Hartley writes the Wild Life column in the Spectator. *His first book was* The Zanzibar Chest; *his book about ranching in northern Kenya will be published in early 2020 by Grove Atlantic.*

droughts. I rate myself not in how I respond to wet seasons, when pastures are as green as Ireland and any fool can come by success, but in the ways we tackle the very dry years when it hardly rains. Our cattle are entirely grass-fed. You could say we farm grass, not cattle, and everything comes down to conserving bush pasture. Alongside the wildlife that wanders in and out, we figure on stocking one adult cow to fifteen acres of pasture. If there were no wild animals, we would have much more grass and could stock twice as many cattle. That would improve our income, but like cattle, wildlife species also benefit the land.

Zebra will eat rank grass and weeds the cattle spurn. Browsing species nibble at separate layers of the bush: giraffes at the top, elands below that, then gerenuks, then other antelope. Elephants bulldoze trees but also clear bush to make way for grass. In other places where the wildlife is killed off or excluded by fences, I have seen invasions of noxious weeds and massive encroachment of bush that strangles pasture.

> **You could say we farm grass, not cattle. Everything comes down to conserving bush pasture.**

My own belief is that getting rid of wildlife probably brings no benefits, whereas biodiversity here helps sustainable farming of the sort pastoralists have been developing for centuries.

Indigenous traditions of pastoralist husbandry inspired the choices of animals we keep. On our farm, I have bred a stud herd of Borans, zebu-type cattle with humps and dewlaps that are perfectly adapted to our arid conditions. The Boran is an entirely indigenous beef breed, selected a century ago by European ranchers from indigenous African stock whose excellent qualities had been nurtured by pastoralists for fifteen hundred years or more. Our district is also sheep country, and the best local breed is the Dorper – a cross between the Dorset Short Horn ram from the United Kingdom and the Blackhead Persian, an indigenous breed so old you can see the animal depicted on the walls of ancient Egyptian tombs. Together the Boran cow and the Dorper sheep combine pastoralist traditions based both on millennia of survival and on modern European science and selective breeding.

Some farmers in our area also own dromedary camels. These are resilient creatures, relatively easy to keep, though slow to breed. Their milk is low in cholesterol and fats, high in vitamins, and fetches a handsome one dollar per liter on the local market. News of the benefits of camel's milk is spreading and we hear supplies are now being sold in European supermarkets.

Donkeys, I discovered a long time ago, are happy creatures that will perform like small tractors and almost never break down if you treat them kindly. When we first started on

the farm in 2003, we pitched a tent and donkeys carried water in jerry cans from a spring. Later the donkeys carried building sand and cement while we were erecting houses and barns.

To make ends meet, I decided to plow up a portion of the farm's virgin soils to grow crops, potentially a better source of income than cattle. Like some other ranches committed to conservation, I saw sacrificing some of our land to cultivation as a way to conserve the majority of it. In a way it was exciting – we were turning over new ground like the mid-nineteenth-century sodbusters of the Great Plains. At the same time I felt guilty, knowing that crops would displace a spectrum of species from cheetah (seven thousand remain globally) and reticulated giraffe (only nine thousand survive). As a beekeeper with fifty hives, I lay awake at night wondering how I could go over to the dark side of applying pesticides and glyphosate. My compromise was to grow fodder crops without chemicals. Yields were lower, but the cultivated pasture became a home to more game, not less, alongside my cattle, so I believe it enhanced both productivity and biodiversity.

G ROWING FODDER CROPS like hay gives me a commodity to market to my pastoralist neighbors. I also sell them Boran bulls and Dorper rams, and this week I am even negotiating to supply a man with four stallion donkeys he wishes to use to carry water. Supporting education for children in the area and trading do not persuade the young men among the pastoralists to stop raiding my cattle – and why should I be

surprised, since rustling has been an honorable pastime since the days of the ancient Greeks? Hours after his birth, the boy god Mercury stole the bulls of his uncle Apollo, the sun god, and tried to confuse his pursuers by placing the animals' hooves on backwards – but was caught and only got himself out of a scrape by making his uncle a lyre that produced beautiful music. The young warriors from Samburu families that neighbor us today do not seem greatly different.

Many times we have had cattle raids and I have trekked after my stolen animals with our herders, sleeping for several nights on the hard ground next to the raiders' and cattle's tracks. Strange as it may seem, those cold evenings with the farmhands, sharing between us a few

cups of water, have been some of my happiest times as a man. Except for seven cattle that were lost completely, I always managed to track them down and persuade the elders to make the young men return the stock.

Apart from that, young warriors might also occasionally hunt lions to prove their manhood, but at least they do not usually hunt other species of wildlife for meat or ivory or horn. They have traditionally believed hunting for wild creatures is beneath them as honorable men, who should only gain wealth through cattle. And the belief is that cattle are gifts from God, or Ngai, specifically to the Samburu. While drinking cups of milky tea in the cattle camps up the hill from my home, I have often heard stories that the Samburu once inhabited the planet Venus. That was

their home until the surface of that world had become so overgrazed that Ngai made a rope that he slung from Venus to the planet Earth. Down the rope the Samburu people climbed, arriving in a wild Garden of Eden. To help them on their way, Ngai sent herds of cattle down to the new homeland, to be a source of milk, blood, meat, and wealth for the pastoralists evermore in the wilderness.

I find consolation in that creation story, because it tells me that since the arrival of humans in this landscape where our family so enjoyed building up a farm, the destiny of humans and cattle have been entwined. But there's a warning there, too: we must care in this world for nature, the pasture, and the wildlife, in order to avoid becoming another overgrazed and denuded planet. ⤳

The Dead Breed Beauty

Today, a falcon cruising above the river.
Dippers, pure grey, about the size of a dove,
splashing in and out of the water in bursts,
doing knee bends as they feed at the shoreline.
Bush tits in the brush, dull and hyperactive.
Farther in, the birds disappear, the canopy
crowds light into odd corners, and life
becomes a rustle of leaves at my feet.

In one of the campsites, a very old spruce
split by lightning halfway up but still living –
from the black scar downward a slug of bark
the size of a Jeep has fallen on three points,
a swooning tripod sculpture half buried
in muck. Campers offer their customary
reverence, cutting chunks for tinder.
Bright tents swell like fungi on the moss.

Strange the human-like gestures of dying
trees, clinging epiphytes, vines along
the riverbank. A cottonwood, roots exposed,
dances; branches laden with epiphytes
suggest women drying their hair. Here
everything climbs at someone's expense
and the dead breed beauty. We savor it
like tramps around a barrel of fire.

RICHARD SPILMAN

ELIZABETH GENOVISE

LEVEL

Photograph by Paul Gilmore

The carpenter's son is dying.

With his boy confined to the blue bedroom at the northwest corner of their ranch on the Piney River, he's moved his workshop out of the garage and onto the patch of sun-dappled grass that slopes down from the child's window to the water. Everything is tilted and precarious, from his hand-hewn worktable to the power tools cooling atop the maple stumps at the water's edge, but it's the only way the boy can watch him work. The carpenter has studied the scene from the blue bedroom window and he understands its appeal: the sun glints off the table saw's blade and heats up the cedar slabs' natural blush, and gold and scarlet leaves lie like daubs of paint on the lawn. Behind it all, the soap-green waters of the Piney River swirl around glossy boulders and divide at the little island of birch trees where the two of them used to sit and fish in the early mornings.

He's a finish carpenter by day, a woodworker by night, and while his working hours have not actually changed since the doctors relinquished his son and said *better you should spend these last days at home*, there's a sense of the days lengthening out between six in the morning and four in the afternoon, in contrast with the impossible speed of twilight's descent. It is the first week of November, and the carpenter is running out of time.

His wife, who left them eighteen months ago, has returned to care for the boy by day. It all enrages him – her returning only now, for this, after everything she did; her cot set up in the third bedroom; the sound of her percolating the coffee and working the toaster as he lathers up in the bathroom before his workday; the

Elizabeth Genovise is an O. Henry Prize recipient and the author of several short story collections. Her third collection is due out from Texas Review Press in May 2019.

fact that he must go to his workday at all, laying trim in strangers' homes as his son diminishes. But there must be money. Money keeps the boy comfortable. Money means he won't lose this house, the only structure on earth to have sheltered and known his child. He'd built the place himself, drawing and redrawing the blueprints for months beforehand. He'd measured everything three times. He'd insured it against every incident, insulated it against all the forces of nature. There wasn't a single nail that hadn't been driven in with all the force of his passion, and it was like the world beginning all over again the day he completed it and walked his wife and infant son through the front door. He was in love with God in those years, with himself too, and when the three of them passed over the threshold and into the living room, he swore he heard the velvet whisper of a window sliding open, as though a passageway between their world and that other dimension had unsealed itself.

But that was ten years ago, and the carpenter is no longer in love with God. In the hour before twilight when his son likes to watch him work, he does not let his face show the fury he feels. He does not let his lips move when he argues with the divine. These arguments are long and convoluted and make him feel like an actor alone on an enormous stage, struggling to play every part, for he must speak God's lines too. The effort exhausts him. He carves away at the cedar, creating towers of red and cream curlicues that inevitably spill riverward with the force of gravity, and he argues, often through the entire hour.

Sometimes there is no back-and-forth. Sometimes he simply says, *You bastard.*

His son has always loved wood, wanted to work it since he was five. It was the greatest pride of the carpenter's life to show him the trees' secrets, the violet in walnut, the rainbows of blue and green concealed in poplar, the whimsical constellations etched into birdseye maple. Cutting western cedar, they would stop to breathe in its deep, floral scent, and the boy would say, *it's like fall leaves on fire.* He'd loved to organize drill bits in their plastic drawers and to uncoil the orange extension cords that lived on their garage floor. He loved the squared-off eraserless pencils his father kept in his jeans pocket and he loved the pop of nail guns, the rumble and gasp of compressors. As a baby his favorite toy had been the carpenter's yellow level with its watery windows and the lone bubble that shifted back and forth, right to left, until centered. Together they would set the level atop dressers, tables, boxes to test the rightness of everything. When the bubble wobbled, then stilled between the two center lines, there was a sense of supreme accomplishment: it was they, the builders, who had properly aligned the world.

The carpenter explained to him that the best pieces were seamless, fluid. He taught him the art of mortise and tenon as if it were a holy language and he the last man to learn it. When they worked with a burl or a live-edged slab, they waited for the tree to tell them what it wished to be. *Listen to the wood. Follow it.* He believed he could teach his son a kind of art, a secret knowledge that would guide him into a profession that would satisfy him beyond any other.

But the boy will create nothing, will become nothing. Every day is the same now: the carpenter pushes the old truck down I-40 as fast as it will go, smokes a cigarette over the course of the last ten miles of his drive, and pulls into the garage. He takes his tools and the cedar out back and sets everything up, creating a soft cadence of sound that he knows will wake his son out of his drugged sleep. Then he goes inside to see his boy. His wife busies herself in another part of the house while he talks with him, helps him with what little he's able to eat. Often the boy is too dazed to say much more than *How is it coming*. When it's clear the boy has tired of talk, the carpenter steps outside to work, an almost silent film shot again and again through the frame of the bedroom window. At twilight he returns to the bedroom to hold his son's hand as he drifts into sleep. He eyes the Star Wars figurines lined up on the windowsill, notes the way the dying light sparks against the blue of Luke Skywalker's lightsaber. It wasn't so long ago that his son was in that familiar stage in which he imagined his own future heroism, the call that would come to him from distant suns, the good he would do armed with a saber or hammer of his own. *You bastard.*

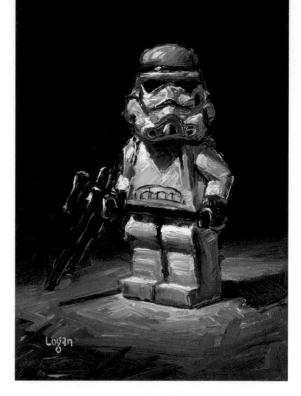

Tonight, the carpenter listens to his house. Is it that the wind has become more ferocious, or is the house less of a bulwark than he'd thought? It seems that the creaks are louder, the whistling keener. To distract himself he thinks of his wife two rooms over, tossing on her cot, and of how much he'd like to turn her out on the street and deny her this protracted goodbye to a child she'd decided wasn't worth her time eighteen months ago. Rage keeps him awake, and he revels in it. He hates them all: his wife, the doctors, God. Sleep won't change that. Even when he does sleep, he wakes enervated from endless dreams of battle. In his dreams the other man is much larger than he is, backlit and shimmering, and always they grapple in a vast, sloping field. Once, he felt himself winning, and then the other man's hand reached out through a shaft of amber light and tapped his thigh. The carpenter collapsed face-down in the soil. Now he recalls with humiliation having walked bleary-eyed into their minister's office the following morning to ask about the story of Jacob wrestling the angel.

"You believe you were wrestling God?" the reverend asked, hand against his chin.

"I just want to know what the story means," the carpenter said, digging his heels into the carpet like a mule. "I read it again this morning and it still doesn't make sense to me."

"To be quite honest with you, I've never understood it myself," the minister admitted.

"It seems like neither of them won."

"That's how it reads to me." A pause, then, "How is he?"

"He's dying, Reverend. What do you want me to say?"

When his phone blares out its five-thirty alarm, he's gummy-eyed and aching, crescent moons carved into his palms from fists clenched in his half-sleep. He'll have to catch whatever shuteye he can on his lunch break so that when he comes home again to work on the cedar, he will not cut himself or topple down the lawn's slope into the stream.

He finds his wife in the kitchen, stirring

his idea. Is that true?"

"Of course it's true. He never lies." Something catches in his voice, and he moves quickly to the front door, zipping up his coat. "Make sure he drinks his water. You have to keep on him." As if it would make any difference.

He works through dinner in a cold wind that whips along the bend of the Piney River, frothing up the rapids. His boy is sitting up in the window – always it fills him with a rush of hope, followed by furious despair, to see that gaunt little body upright amidst the tangled blankets, the bald head like a question mark – and the pressure is on. He carves the cedar, stopping only to wipe snot from his nose and dewy water from the corners of his eyes. "It will sound just like the river," he calls out; the boy's window is cracked just half an inch. "It'll flow in perfect lines. You'll see."

His son nods. His confidence in his father is absolute.

The carpenter adjusts the shims under two legs of the worktable, something he must constantly do on the hillock. When he returns to his carving, he bites his lower lip in concentration, shaping the cedar. *The problem, God, is that you think you know what this is like, but you don't.*

– Don't I?

– Your son chose to die. And he was no child. He was thirty-three years old.

– Do you really think that made it any easier for me?

– Apparently not, as you simply sat there and watched.

– You don't know what I did.

– Let me put it to you another way. There was purpose in what happened to your son, at least that's what you claim. There is no purpose to this. None.

a skillet of scrambled eggs and sausage. She jumps a little at the sight of him. "I thought you'd want something better than toast for a change," she says, not quite looking at him. The smell of the sausage, maple and spice, has the same effect on him as the smell of her vanilla perfume – it's too familiar, too officious.

He reaches for his Carhartt. "I don't have time. The faster I get there, the faster I get home."

Her eyes are circled with gray, her hair bound in a limp braid. She's wearing a well-worn lilac robe he recognizes from early in their marriage, a silken thing trimmed with deep violet. He'd bought it for her as an apology gift after forgetting their anniversary. She's about to say something else when the words fly from his mouth: "Haven't seen that one in a while. It's kind of sexy for around the house, don't you think? But maybe you're used to that."

She closes her eyes briefly before turning back to the stove and shutting off the burner. "I'll just make you some toast."

"I'll get something on the way."

"He won't tell me, you know." She says this quietly, her back still turned. "He won't say what it is he asked you to make. What you're working on out there. I thought maybe a table – I saw that slab with the curvy edge that you've got up against the side of the house. But he said it wasn't just a table, and that the whole thing was

Silence. No cue card.

He carves the cedar, breathing in the scent of leaves on fire. His strokes reveal the deep violet-red that always drew his son's eye. When finished, this thing will be precisely what they've wished it to be, God's senseless interference in their lives notwithstanding.

On a Tuesday night, the boy asks for the carpenter's cigarettes. In his complete surprise, he hands over the wrinkled pack of Camels and waits. The boy turns them over in his hands. Not trusting his voice, the carpenter asks, "Did you want to try one, son? Is that it?" It's too horrible, the idea that his ten-year-old might have something so dreary and mundane on his bucket list, but trips to the ocean or hot air balloon rides are out of the question, and so there's a part of him that's perfectly willing to take out his lighter and hold it for the boy until his thumb burns.

His son breaks into a weak laugh. Oh, that sound – it's a broken bell in an abandoned cathedral. "No, Dad," he says, covering his mouth. It's something he only began to do after his diagnosis, as though afraid he'd infect somebody with his own mortality. "No." He mashes the carton between his hands, twisting it and mangling it until the cigarettes are unusable. The carpenter watches this in silence.

"I want you to stop," his boy says, glassy-eyed with the mess in his lap. "I know it's hard to quit, but stop. I want you to live a long –"

"I'll stop," he says, too loudly. "Here." He takes the ruined cigarettes and dumps them in the plastic trashcan by the bed. His hands are shaking. "Now get some sleep. Or did you want to read more Narnia tonight?"

They read together. It is something the carpenter used to work hard at dodging even when his son was tiny; he'd tell the boy he needed to get to bed early when in truth it was simply that he hated to read aloud and hated the way he stumbled over words his son could already pronounce with ease. Now he's unable to say no to any request and so he works his way through the story, licking the calloused tip of his index finger to turn the pages. The chapter finished, he takes his shower, where he weeps and bangs his fist against the grey tile.

When he emerges from the bathroom it seems to him that the floor has tilted; he stumbles over his own feet as he crosses the hall toward his room. His wife, also on her way to bed, stops in the hallway and reaches out to touch his arm. He steps back. In the moonlight, she's pale and wide-eyed, thin in her nightgown. "I heard you," she says softly. Her voice is unbearable, full of the old tenderness. "Please, talk to me. Let me listen."

"There's nothing to say, and there's not going to be. Just help me care for him, and then go back to your life. I know you want to do right by him, so do it. Then let me be."

She says his name, her hands in her hair.

"I have to get some sleep. You should, too."

He is too tired and cried-out for anger, but when he lies back in his bed, he finds himself addressing God again: *I know there's no blood, no whip, and no nails. But I believe my son has suffered worse than yours.*

Silence. He tries again: *At least yours was quick. One day.*

– One day is an eternity here.

– At least your wife didn't betray you.

– Every woman on earth has betrayed me at one point or another, as has every man.

– But did you have to walk around your house with them all there as you watched your boy bleed out? Did you have to pretend to believe they cared?

– There was nothing to pretend. I knew they cared, or that they would, eventually.

– I'm not willing to wait for eventually.

– Then it's a good thing I am.

They grapple again in the carpenter's dreams, God's skin hot and glimmering in his hands as they tussle in the field.

The first three tiers are done, as are the live-edged chutes braided between them like vines. His son's inspiration came from a video of a Rube Goldberg device, watched on a laptop balanced across his knees in bed back when he had the energy to do such things. The carpenter hadn't understood why it pleased him so, but he knew it was his last chance to be the boy's hands, serve as a proxy to his imagination. Somehow he also knew that this object, whatever it turned out to be, would be his life raft after the boy passed. It would be the last thing to embody the boy's spirit. It had to be perfect.

It's an unusually quiet Thursday, the wind silent and the house still, no cars on the byway. His wife has gone to do the shopping and whatever else she does in town. His son is already asleep at six-thirty in the evening. The sycamores and birches are ghostly in the falling light. The carpenter works on, sweating in the November chill, hoping to finish the fourth tier before darkness falls. One of the boy's doctors warned him that if the boy began to talk a good deal, or make a sudden show of energy or agitation, it meant he was close. There's been no sign of such feverish enthusiasm, but earlier, when the carpenter came home, he found the boy talking with his mother, the words coming fast, hurling over one another in their rush to some unknown end. "You remember them, don't you? Those water games at Uncle Andy's, with the rubber buttons and the plastic windows. You could hook rings on poles or try to get basketballs into a net. You never told me I was too old to play with them. You never said anything when I went to ask for them. You

knew I just loved them." His mother was crying weakly through this strange speech, but the boy didn't seem to notice; his face was aglow, his eyes faraway. "I just wanted to say thanks for that. I never did."

He has to fight to slow his hands so as not to make a mistake. He can't afford the time it will take to redo anything.

We were happy then, God, during the Uncle Andy's days. You must have been there with us at that craphole of a pancake restaurant at least once. It's closed, you know.

– I remember. And yes, I know it's closed.

– But the truth is she never loved me. She walked right out on us.

– Why?

– Because she was a whore, that's why. She saw something she wanted and just went after it.

– So it was all her, then.

– Excuse me?

– There was nothing you did to drive her away. You made no mistakes in your marriage.

– That's right.

Silence. Then he lets out a yelp; he's driven a one-inch splinter into the skin of his palm, just below his ring finger. "Shit," he whispers. He hasn't sworn out loud in months. The splinter is deep, and it hurts; he can't flex his fingers without feeling it edge up against the nerves.

He marches inside, into the half-bath off the laundry room where he keeps his first aid kit. A minute later his wife is in the doorway, mouth pursed in concern. "Splinter?" she asks.

When they were young, she used to pull them out for him with tweezers; in fact, she'd had a talent for it, sliding the needles of wood out with a grace that convinced him she was born to be a mother. Now she moves close to him and studies his palm beneath the bathroom's yellow light. "That's a bad one," she affirms. "Let me try?"

"I've got it."

"Don't be stupid, you'll just make it worse. Christ, what are you doing?"

He's using a pin to rip up the skin and expose the splinter's frayed end. Tiny beads of blood bloom from the wound.

"Stop," she says, grabbing his hand. "Do you have tweezers in here? Good. Come on. Just hold your hand open."

Grinding his teeth, he holds still as she gently pushes the splinter toward herself, pressing her nail firmly against the opposite end. The bit that emerges is infinitesimally small, barely visible to the carpenter's eye, but his wife deftly nips it between the tweezer's paws and has drawn it out before he can breathe. With a small laugh, she deposits the shard on the edge of the sink. "Put it under your pillow for the splinter fairy," she says. It's an old joke of theirs, and he almost smiles before he catches himself.

"Thanks," he says, putting the first aid kit back under the sink. "I have to get back."

"I think you'd better stay in." She hesitates. "He's coming in and out of sleep, but when he's out, he's talking."

"I know. I heard a little, earlier. But I have to get this done. You understand?"

Another pause. "Yes. I think so. But when it's dark – I'd like to be with you two, together, if you don't mind. I hate this separation at night. I hate it."

"You hate the separation." He says this, then looks deliberately into her eyes.

She flinches but holds her ground. "Yes, I do."

"Do what you want," he hears himself say.

"Thank you. I'd like to sit with you all tonight."

He nods once and returns to the backyard to attach the cedar slab tabletop.

t's been years since he was vulnerable to splinters. Over time he's grown a second skin over his hands, a tough whitened shell pocked with callouses. His cuticles are ragged but

healthy, his nails thick and pink. It unsettles him to find that there are soft spots in the moonlike surface of his palms. He works steadily but more carefully over the course of the next three evenings, wary of the chisel's path.

On the third night he allows his wife again to sit with him in his boy's bedroom. Suddenly finished with reading, the boy seems to want only to talk, and there is no recognizable pattern to his thoughts. As the nebulous grain of burl wood sends power tools cartwheeling in all directions and thus requires the slow and careful use of hand tools, so his son's swirling mind requires more attention and naked intuition than the carpenter is used to having demanded of him.

"I stole from your purse once," the boy tells his mother, mouth twisting.

"It's all right," his mother says, patting his hand. "It doesn't matter."

"There was a kid at school everyone made fun of. I liked her but I joined in anyway," their boy goes on, eyes on the ceiling. "Her name was Ashley and I called her Ash-can like everyone else. Her mom was an alcoholic and all her clothes were Salvation Army, that's why people ripped on her. That was just last year."

"Do you want some water, son?" the carpenter asks, leaning forward.

"I had these mean thoughts. Just mean," he continues, his voice fading a little.

"Everyone does," his mother says.

Now the boy's eyes are filling up. "I kicked a bird in the street. It wasn't dead yet but I kicked it just like everyone else was. Why did I do that?"

"Baby, baby. Come on now."

The carpenter says, "Why are you telling us all this, son?" Then, unconsciously echoing his wife, "Come on, now."

"I don't know." He looks at his father, and his hazel eyes are piercing, more complex in their whorls of color than he can recall them being. What he sees there is all the astounding intricacy of walnut – hazel, palest blue, violet, sienna. How had he never noticed before? The boy says, "I want to talk to some people and tell them sorry but I'm – I'm too tired." He lets his head fall back on the mountain of pillows behind him. His skin is ghastly white, rivers of capillaries showing through. "I'm too tired."

The carpenter rises. "You get some sleep, buddy. Getting close on the waterfall, you know."

"I know."

Alone in the backyard beneath the stars, he works his jaw, the fury rising in his blood like a fever. *You are sick, do you know that.*

– I beg your pardon.

– You're in control, you can put an end to it. Make him well. At the very least don't make me sit here and watch him beat himself up for these tiny little sins.

– He's a child with the heart of a man. He's taking responsibility for whatever he can.

– You want me to say he's a better man than I am? Fine. He's always been the better man, even at eight years old. He had more patience. He thinks I'd use mortise and tenon every time, but I like the power tools. It was him that changed me. He walked into the garage at three years old and all of a sudden I never wanted to see a hinge or a screw sticking out of the wood again.

– Now you're saying something real, son.

– Don't call me that. Don't cross the line.

– I'm like your wife. I want to sit in the room with you at night.

– You can both keep to your own rooms, thanks very much. Make him well again and I'll let you into my house whenever you ask.

– Those are the conditions?

– Yes.

– I'm afraid I –

– That's what I thought.

He pulls the green tarpaulin off the almost-finished product and studies it in a wash of moonlight. When he replaces the tarp and steps back, he miscalculates the ground's slope and almost falls backward into the river. Cursing, he steadies himself and glances furtively at his son's window. The room is dark. After a long pause filled with only his own breathing, he crouches down and drives his fist into the soft earth, punishing it for its unevenness, its unsuitability. This little hillock was just a pretty place to read and watch the water back when their lives were simple. Now it's another body to wrestle, a force continually pushing back. Even the trees' browning foliage and the river-fed undergrowth seem to be encroaching on his workspace with malicious intent. He wants his garage with its level concrete floor and walls of neatly organized tools; he wants its insulated quiet and its humid air smelling of pine sap and mineral oil. He wants his son to wander in after school in his blue jeans with his hair all cowlicky, eating a Twizzler or slurping a Dr. Pepper the way he used to, and he wants the God he grew up with, who never asked questions or demanded much of anything save that he do his job, put food on the table, and attend church when he could to ask for what he wanted.

His son rarely wants the television, but today he asks for a movie – *Ben-Hur*, a film he'd once watched obsessively, two or three times a week the year he turned eight. The carpenter finds the old VHS in a jumble of tapes in the living room closet and cleans it off. Then he wheels in the metal tool cart he converted into a portable TV stand awhile back, when the boy became too weak to leave his bed. "I'd watch it with you, but I have to work," he says, pointing out the window at his worktable.

"I know. Can you put it in now?"

"Sure." He slides in the tape, grateful for the reprieve. He never could bear to watch this movie with his son. It was the way his boy always stood at wide-eyed attention for those quiet scenes in which Christ materialized. Always the man was seen only from behind, as when he offered water to the disconsolate Ben-Hur in the desert, but the music shifted dramatically when he appeared. It was the sound of water moving in a slow cascade down a staircase of leaves, and it played at the film's end, too, when the lepers were healed in the rains of the thunderstorm that followed Christ's death. The carpenter says, "Are you sure you want to watch this?"

His boy looks at him. "Is it OK? I won't if you don't want me to."

"Yes, yes. Of course it's OK. I'll be right there, right outside." He exits the room, fists opening and closing at his side. He will surely wrestle God again tonight, and try to pull the other down the sloping field into whatever ditch lies below.

He works through a downpour the following night, beneath a makeshift shanty constructed of tarp. He returns from the yard to find most of the house in inky darkness save for the faint glow of his son's nightlight and a square of light from the entrance to the basement. There's not much down there besides canned food, the furnace, and the boy's bicycle, and he starts down the stairs thinking he's left the light on from earlier when he went in search of tomato soup. He finds his wife sitting in the middle of the gritty brown carpet between the storage closet and the bike rack, a photo album open across her lap.

"You shouldn't," the carpenter says roughly.

His wife snaps the book closed. Rising, she settles it back into the box she'd pulled out from under the old picnic bench they once used as a laundry table. "I know. I don't have any right to look. I'm sorry." She moves past him to the stairs, biting the inside of her cheek.

What he'd meant was that it would only hurt her to look at those pictures of their son when he was healthy and carefree. It stung him, what she'd assumed; just what sort of person did she think he was? He stands there a minute with his hands jammed into his pockets and then he strides over to the box and removes the photo album. He sits down on the picnic bench and opens the book to a random page. The crinkle of cellophane is enough to make his chest tighten; it's the sound of an intact family, of a woman scrapbooking at a kitchen table while her husband fixes the dryer and her son builds action-figure kingdoms out of blankets and chairs. It takes everything he has to keep paging through.

It could be that the lighting down here is off, or that he's bone-tired. But the photographs he's on look completely foreign to him, as though another man, someone he didn't particularly like, had stepped in for a cameo, then ducked back out just after the pictures were taken. He squints. Here is a snapshot of the two of them in his garage workshop. The carpenter is bent over the worktable, focused and intent; the son eyes him from the table's end, exactly like a dog awaiting orders. Is it his imagination, or does he, the carpenter, look a little too impressed with himself? In another, they're all at the aquarium; his boy is clutching a stuffed orca whale to his chest, and just behind him, the carpenter is rolling his eyes. He remembers that day and what he'd said to his son: "You're too old, and it's a girl's toy anyway." Why had he said that? He closes the album and begins to pace the basement floor. In a moment, it comes to him, the fight he'd had with his son not long before that trip.

Together they'd watched a documentary film on the captivity of orca whales. The orcas were hunted, separated from their little ones, and kept in tiny pools that were little more than bathtubs. Their dorsal fins flopped over in defeat after so many years of swimming in circles. They made desperate attempts to contact their loved ones via long-range vocals that went unanswered. The carpenter's son wept at the film's conclusion. "It won't go on forever," he'd tried to tell his son. "Come on, now. There are even good parts to keeping those whales in captivity. They learn about them, so they can help the ones in the wild if they get sick." He had no idea if this was true. His son looked up and said, "I'm thinking about what they said about how hard the whales work to make the trainers happy. How they try to do all the tricks. I think I know why they do that."

"What?"

"They do it because they think if they do it good enough, the trainers will love them.

And if they love them, maybe they'll let them go free." The boy dissolved into fresh tears, completely breaking down, and then the carpenter's wife rushed in, wanting to know what had happened.

Still pacing the chilled basement, the carpenter has the monstrous thought that his boy had looked at him in the same way. Was he really in love with the wood and the workshop, or did he feel that he had to perform those tricks in order to win his father's love?

Instead of going to bed, he lets himself back out of the house and down to the river. *I was always interested in him. I let him be his own person.*

– How many times did you cut him off when he tried to talk to you, unless of course he had a hammer in his hand?

– He loved the workshop. I know that was real.

– Yes. He loved books too, and riding through Spring City on an autumn day, kicking up leaves with his tires. He loved that girl he tried to tell you about when he was eight. He loved to write little poems.

– I'm speaking for you, God, so if you know all this, it means I know it too.

– Please. You were the same way with your wife. No curiosity.

– No curiosity? What the hell?

– Language. You don't remember that strange little argument you had, years ago? She left her diary open and was surprised that you never touched it or even asked about it. "I'd have to read yours," she said, and you laughed and said, "Why?" What were you so afraid of?

– This is insane. I knew them. I was always there for them. Do you see this house I built? Do you know what went into it?

– Yes. You built the house.

In the silence that follows, the carpenter kicks at a tool bucket he's left out on the grass.

The minister arrives on a Sunday evening. It is not a social visit – the carpenter has made that clear. It's business. The minister's brother is a gifted metalworker and the carpenter has put in a request. Per the carpenter's instructions, the minister comes straight to the backyard with the package, bypassing the house.

"About time," the carpenter says. "He made it exactly to my specifications?"

"He says he did. I don't know anything about it, so you have to trust him. He did comment on just how exact those specifications were." The minister says this with a small smile that infuriates the carpenter.

"It has to be right. This isn't a game."

"I know that. Will you open it now? I'd like to see it myself."

Begrudgingly he opens the package. What the metalworker has made is a kind of net of copper leaves, bound together at the stems; they glimmer in the November sun, throwing off sparks of red and orange. The carpenter moves his fingers over one of the leaves and nods. Almost forgetting the minister, he pulls a sheet off his nearly-finished table and kneels at its base.

The minister gasps.

"Wait," the carpenter murmurs, distracted. The network of leaves will have to be properly attached later, but he'll be able to tell right away if it was made correctly. He holds it up against the descending tiers of cedar, the delicate live-edged shelves that form a kind of staircase below the tabletop. The minister's brother has been true. It's a flawless fit, the leaves clinging to the tiers like hands, forming cups below the shelves' tilted ends, creating the illusion of movement and breath.

"Never in my life," the minister says softly. "I've never seen anything like this."

The carpenter hesitates. "It was my boy's idea. The whole design."

"Am I right, that what you need finally is water to –"

Sharply he says, "Don't spoil it. Don't say it." It occurs to him that what he's been working on is as close to something sacred as he's ever gotten. "There's something wrong in saying it out loud."

"I understand. I think all artists feel that way. My brother –"

"Tell him he's done well. Here." He digs into his back pocket for the check he's had written out for days.

"Michael won't take pay for this," the minister says gently, holding up both hands.

"I know what it costs to work with copper. I've been putting this aside since the day my boy told me his idea. Give it to him." He thrusts the check into the minister's shirt pocket and turns away. "I'm close," he says. "I have to finish."

"I'll let you be, then." The minister hovers a second. "We're all praying for your family, you know."

"Thanks," he says shortly. "Thanks a million."

The minister gone, he tilts his head back to study the trees, whose leaves are now almost completely brown. *You want me to say it, God? Fine. I wanted to tell him it was my idea. I wanted the credit.*

– Why?

– I looked pretty full of myself in those

workshop pictures.

– You don't love the wood?

– I do. But I came to misuse it. As I misused my son.

– How so?

– I made the wood his only route to my love. I was afraid he'd grow up into someone I couldn't understand. Now you see what a repulsive son of a bitch I am.

– You did not misuse your son. You tried to keep him close. And you cared for him all on your own after your wife left.

– I did wrong by her, too. It's time I took it like a man, like he does.

Silence.

– What do I do?

– Finish what you began.

He tinkers with the leaf network for five or ten minutes before circling around to the garage for some tools. He's rummaging through a plastic drawer when he hears voices in the house – his wife's and the minister's. He closes the drawer, heels off his boots so he can let himself into the laundry room without alerting them to his presence.

They're in the kitchen, out of sight, and he wants to go in to them, but something pins him in place long enough to hear his wife say, "Isn't that in the Bible? The sins of the parents get visited on the sons? It's a punishment for what I did to them."

"Don't think of it as a punishment." The minister hesitates; the carpenter, standing beside the dryer, holds his breath. "Think of it as an awakening."

The carpenter backs quietly into the garage. He's forgotten what tool he needed, and he stands there awhile, hands on his head like a man awaiting arrest.

They have a long and difficult Monday night during which the boy keeps climbing out of bed, mumbling unintelligibly, and they have to keep leading him back. On the road coming home from work Tuesday, the carpenter is too dazed to realize he's going fifteen miles over the speed limit. The officer who pulls him over looks closely at him and asks too many questions. "You can just give me the ticket," the carpenter says finally, to end it. The other man's look of fatherly concern is insufferable.

When he comes home, ticket in hand, he finds his wife collapsed at the kitchen table, crying stormily into her hands. His immediate reaction is to throw down his lunch sack and rush down the hall to the boy's room. Heart slamming in his chest, he shoulders open the door. But his son is still alive, still breathing, sleeping with his arms strewn behind his head like a baby's. The window's curtains are pulled back as usual to reveal the still life of the outdoor workshop.

Back in the kitchen, he flicks the overhead light on and off to get his wife to look up at him. "You scared the living hell out of me," he says breathlessly. "You know that?" Then,

"What happened?"

She looks at him from between the slats of her fingers. "He said he forgives me. For leaving."

There is a long pause before he lowers himself heavily into the chair across from her.

"I brought it up. I wanted him to know – I needed him to know it wasn't him I'd left. It wasn't anything about him. I told him it was just me, doing something selfish. Being an idiot. And he said it didn't matter." She lets out a sound somewhere between a laugh and a sob. "He said, 'It wasn't very long, in the scheme of things.' Where'd he get a phrase like that? I barely kept it together until he went to sleep. Then I came in here." From below the table she lifts a bottle of red wine, half empty. "I had this in my car. A couple of bottles, actually. I've been going at it some nights, when I don't know if I can take it."

"I do that too. I mean I did. After you left." He rubs his hardened fingertip over a scratch in the tabletop, no doubt made by his son's pen or colored pencil years ago.

"I was horrible to you."

He says nothing.

"I know you think it was just about him –"

"Don't say it –"

"But it wasn't. I know that now. I was never even in love, and it wasn't about the sex either. I was in love with his loving me. I was this whole new person with him. I got to pretend to have no history. You understand? It was all lies, me thinking I could just start over and be perfect."

Quietly he says, "What all had you done that you wanted so bad to be someone else?"

"It wasn't any one big thing. It was all the little things. I wasn't the person I'd planned on being. I'd just – disappointed myself. I just didn't like who I was."

"You can say it, you know. I'm sure if you felt like a disappointment, it was because I treated you like one." Stunned at having said

it, he sits back in his chair. He can't quite bring himself to look at her.

From the boy's room comes a rustling, and then his faint voice: "Dad?"

The carpenter leaps up from the table. "Son?"

In the blue bedroom, his boy is standing beside the window, and he's as thin and spectral there as the birches on the river. "You're not going to work on it tonight?" he asks, the words warbly.

"Yes, yes, of course I am. I just got home a little late. I'll get out there right now, unless you don't want me to."

"I do want you to. Can you use the floodlights like we used to? I want to sleep to the floodlights."

"Of course. But you need to be in bed, kiddo. Come on."

He works into the night, blinking through the glare. Dust and flying woodchips look like moths in the white light and in his peripheral vision always he sees his son's shape, propped up by the pillows and bathed in the soft radiance of the maple veneer nightlight he keeps plugged in beside the bed.

If you had killed him tonight, I'd never have forgiven you.

The fountain table is nearly complete. All that's needed are some finishing touches, and then the water. Two or three more days. He calls in sick to work before he falls into bed, his right hand still fitted around the ghost of the chisel.

The last Friday of November, the carpenter finishes the table just after dark. In his exhaustion he falls asleep at eight o'clock and dreams not of wrestling gods but of his wife and young son. It's winter in this memory, and they're in a warmly-lit café in Spring City's little downtown. Snow is falling slowly past the windows, catching up the red and green of Christmas lights strung across the glass. His son sips cocoa and walks over to the café's chalkboard to write something. A poem, a haiku of sorts. Why hadn't the carpenter written it down on a napkin, the back of a receipt, anything? His wife smiles over her tea and reaches for his hand. In the reality of the memory, he stood up to get himself a refill; they'd been fighting over who knew what. But in the dream, he entwines his fingers through hers and squeezes. They have a brilliant child, an empath with a gift for words and a love of beauty. He'll be an artist or writer or filmmaker someday, or perhaps a preacher. And the two of them – they have a long and winding road to walk together. They'll start tonight. He'll ask her . . . he'll turn over in bed, and he'll ask her. . . .

He wakes in a cold sweat. He can hear his son's voice, avid, the words a babbling brook. He leaps out of bed and bolts down the hall.

His boy is lying back against the pillows, hands working above his chest. His eyes are circles and there is an ethereal light there the carpenter has never seen, even in the boy's worst fevers. "I knew I'd see you again," he says when his father leans over him and touches his cheek. "I knew it, I knew it, I knew it . . . ."

The carpenter calls his wife's name.

The boy is still talking as the two of them sit down on the bed beside him. "It's the same sound," he says, shaking his head in wonder. "It's just the same."

His mother covers her eyes, then pulls her hand down to look at the carpenter. "He's going," she says softly.

"No." The carpenter rises, looks frantically out the window. "He has to see it. I can turn on the floodlights. I finished it tonight – all I need is water –"

"Don't you leave." Her voice is fierce, her hand on his wrist. "Don't you dare leave."

His throat is so tight he doesn't think he can respond. Slowly he sits back down on the edge of the bed. Somehow, his wife's hand ends up in his, and he knows he's holding too tight, hurting her, but he can't seem to let go. Their boy looks at their hands and nods. Then, to his father, "It wasn't for me anyway . . . for you –"

The boy's eyes shift to the window. They follow his gaze. The carpenter, now clutching his son's fingers in his left hand, sees moonlight on the rapids, the white arms of sycamores, the silhouettes of his worktable and tools and the fountain table beneath its shroud. When he looks back, his boy has gone.

It is at dawn that he goes into the frosted backyard alone and lifts the sheet from the fountain table. Down at the river he fills a metal pail with icy water and hauls it up the slope of the lawn. He holds the bucket in his arms, feeling its weight. Then, slowly, he pours the silver water down the first shelf of cedar. It spills down in bright rivulets, rushing right, then left, then right again, cascading down the cedar steps and branching off the copper leaves to form tiny waterfalls that rejoin the first river in its voyage toward the table's base. The cedar's deep violets and reds fire to life like old leaves drawing new breath. The new sun refracts off the water just as the carpenter had hoped it would; sparks fly and the copper of the leaves trembles as trout tremble within a current.

The stream is moving fast, but just before it spills out onto the soil, the carpenter drops to the ground and lays his head at the table's base. Frigid water bathes his neck and hair. It shocks him, but he stays where he is until the last drop has soaked into his clothes.

He rises and staggers back into the house. There is much to do today, and he can only think so far ahead. But he will ask his wife to stay with him tonight in their bed. He'll sleep beside her, and he'll sleep beside his God, level at last. ➤

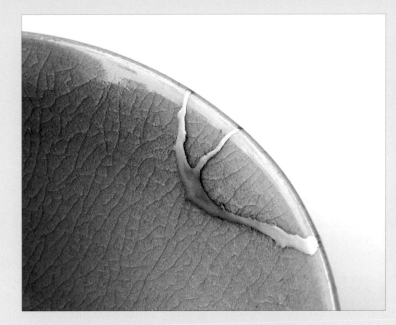

Kintsugi by
Natsuyo
Watanabe

Cloth and Cup

Sometimes a life seems to be
working so well. And then
something tears, and even
though mended, wearable,
is flawed, a scar in the cloth.

Then there's the cup with
the stained crack. Your favorite.
Unique, a tender blue, hand-made on
a wheel that soon enough slowed to a finish.

Fired forty years ago, it still
works. Every day you pour water on
the teabag, careful to avoid
a scalding spill, adding honey.

As you drink, (a chip in the rounded
edge against your lip), you ponder
your fondness for it. Like you, it's
a survivor, flawed but familiar.

The wheel is still there, gathering dust.

LUCI SHAW ➤

Editors' Picks

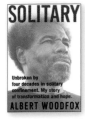

Solitary
Albert Woodfox (Grove Press)

Angola State Penitentiary, Louisiana's sprawling 18,000-acre prison, was originally a slave plantation. Now it's a modern correctional facility. Or is it?

Plough was recently pitched a book about Angola's transformation under Warden Burl Cain, who championed state-sponsored "rehabilitation through Christ." We turned the proposal down; something smelled rotten. Turns out it wasn't just the stacks of expired canned goods prisoners were relabeling for a private company to resell in Latin America. No, even while the prisoner-led churches thrived under Warden Cain, three unrepentant inmates guilty of what he called "black pantherism" were being held for decades in solitary confinement.

Albert Woodfox grew up poor in New Orleans' Sixth Ward. After his first three-year stint at Angola, he returned there on an armed robbery conviction. There he encountered the Black Panthers, who taught him to respect himself and care about others. Soon Woodfox was arranging to protect new arrivals from rape and organizing prisoners to demand better treatment. When a guard was killed, he and a fellow Panther were falsely charged with the murder and removed to solitary confinement.

Around 2000, Woodfox became a cause célèbre and finally acquired competent legal representation. But it would take another sixteen years to win his freedom. By then he had spent forty-three years in solitary, more than anyone in US history.

It is beyond remarkable that Woodfox emerged "unbroken" by his ordeal. His book is a testimony to endurance and integrity – and a damning indictment of a practice currently affecting some 80,000 US prisoners.

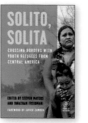

Solito, Solita
Crossing Borders with Youth Refugees from Central America
Edited by J. Freedman and S. Mayers (Haymarket Books)

Certain US politicians claim that child migrants are invading our borders in hordes of thousands. Far more often, though, these children find themselves alone, the *solito* or *solita* of this book's title. The book contains fifteen first-person accounts by asylum seekers from Honduras, El Salvador, and Guatemala, edited from interviews between 2014 and 2018. It sheds light on who these migrants are, why they are fleeing their homes, and what US Americans can do in response.

Some of these stories are haunting. When Julio Zavala was six, his mother kicked him out. He turned to gangs for protection; they sent him to kill his family to prove his loyalty. He fled instead, arriving in the United States at age fourteen. After a brush with prison, he turned his life around. When he called his mother to forgive her, a relative told him gangs had killed her and it was his fault. In April 2018, Julio died of a drug overdose.

Other stories are more hopeful. Many of these youth show a tenacity and maturity reminiscent of an older generation of Americans shaped by the Depression and World Wars. Several, granted asylum, are putting themselves through college while working long hours to support family back home.

The book features many grassroots organizations that help asylum seekers. But, as the editors emphasize, more must be done to address the insecurity in migrants' countries of origin – the problems that caused these children to flee in the first place. ⇥

A pilgrim rests at the Church of the Holy Sepulchre, Jerusalem.

The Necessity of Reverence

Cultivating a pilgrim's eye for beauty

SISTER DOMINIC MARY HEATH

"HOLY SEPULCHRE, BATMAN!" It's a saying my family coined on a trip to the Holy Land when I was a child. The idea of pilgrimage was still foreign to my family of American evangelicals, and I remember my dad turning to me as we entered the tomb of Christ in Jerusalem, a site at the heart of Christian devotion to the Incarnation: "Holy Sepulchre . . . Batman!" he said, and I echoed it back to him.

This saying, which has since become something of a family joke, captures the typical irreverence of tourists. Too often, that's what tourists do: tame uncommon beauty with everyday irreverence. For every one soul flooded with grace in a sacred space like the ancient Church of the Holy Sepulchre, there are floods of tourists moved simply by curiosity and a very natural desire to see something famous. Sudden exposure to

Sister Dominic Mary Heath, OP, is a Dominican nun at Our Lady of Grace Monastery, North Guilford, Connecticut.

beauty just doesn't work miracles for a lot of people, even those who should "know better."

Discussions on beauty in evangelization tend to overlook just this point: conversion is a grace, but reverence is cultivated. Reverence is not a heightened emotional state provoked by a bring-me-to-my-knees aesthetic experience. It's an attitude that involves a certain way of seeing the world, a discernment of the excellence in things and particularly the excellence of God, which gives me the capacity first to recognize, then to receive, the beautiful. Reverence can tell me what I'm looking for when I behold something beautiful. And when it tells me I'm looking for Christ, I become a pilgrim with purpose, not a tourist without aim.

> **Conversion is a grace, but reverence is cultivated.**

The good news is that even if you can't get to a famous cathedral in Jerusalem, Cologne, Santiago, or Westminster, you can cultivate reverence.

What Is Reverence?

Reverence is the attitude of submission we owe to God as the source of our being. It is the posture proper to us as human persons who should recognize our finitude: *I did not make myself and I do not hold myself in being.* Reverence can be called a *posture* in both senses of the word: it involves not only the subjection of our minds, but also the humbling of our bodies. The first leads to the second, like charity in the heart leads to charity in deeds. In fact, the posture of reverence is formed in us by a whole host of virtues with charity at their head. Reverence thrives on charity because charity is a love of God, self, and neighbor for God's own sake that in*forms* all the Christian virtues, animating and perfecting them like the soul in*forms*, or animates, the body.

Charity forms reverence in us, first, by teaching us the fear of the Lord. When our fear of God is a *slave*-like fear of punishment, our reverence is real but imperfect. Reverence reaches its perfection when our fear matures into a *child*like dread of separation from a beloved Father. As Saint Augustine says, "It is one thing to be afraid he may come, another to be afraid he may leave you." Brought to perfection by love, reverence becomes the sign of the Holy Spirit in us crying "Abba! Father!" (Gal. 4:6).

God, of course, is not just any father; he is the first principle of creation and the eternal wisdom governing all things. We revere him not because of his *goodness* (this is the object of our love), but because of his *excellence*. Reverence takes excellence as its object because it is a humbled response to what excels us. This gives reverence a special relation to another set of virtues – the virtues of justice.

Most of us know that justice renders to each what he or she is due. What may be surprising, however, is the name justice takes when it renders due reverence to God: it's called *religion*. Religion is the moral virtue, belonging to justice, that pays God reverence by interior acts of devotion and prayer and by exterior acts of adoration and sacrifice. These acts are virtuous, not when they are sudden and coerced, but when they are sweet and habitual to us. A virtue in this sense is always something we're good at. Religion makes us good at revering God because it gives us the ready will to do so. A person with the habit of religion has the benefit of a soul rightly ordered to God and creation.

Right order is the principle that distinguishes Christian reverence from all merely natural instincts about God as a "higher power." It's no longer possible for us to reverence God generically when we know that the eternal Word, Second Person of the Trinity, has forever united himself to human nature (the ongoing mystery we call the Incarnation). All I just said about reverence as a humbled posture expressing charity, religion, and holy fear applies equally to our reverence for Christ in his humanity – his body, blood, and soul – and divinity. The Incarnation makes one huge exception to the rule that we must worship the Creator of beauty but not the beauty of creation: Christ is simply both.

Christ, in his humanity, brings the excellence of God near to us. We owe him reverence because he is God from eternity, but also because his sacred humanity is the source of our life of grace in time and space. It is for our sakes, according to Saint Cyril of Alexandria, that the Spirit descended upon Christ at his baptism: "If we reason correctly, and use also the testimony of Scripture, we can see that Christ did not receive the Spirit for himself, but rather for us in him; for it is through Christ that all gifts come down to us."

Reverence, rightly ordered, is a humbled submission to the one who humbled himself to share in our humanity, who continues to share his divine life with us. An attitude like this will overflow into our lives: at the Holy Sepulchre it could mean all the difference between Mary Magdalene's encounter with the one who has risen and my own childhood encounter with the one who flies over Gotham.

Imitating Reverence

No matter how old we are, the first way to learn reverence is in the childlike mode of imitation. A second story from my early trip to the Holy Land illustrates this for me vividly. This time we were in Emmaus, the village mentioned at the very end of Luke's Gospel. In that account,

two Christian disciples meet a stranger on their journey from Jerusalem but only recognize him as Christ "in the breaking of the bread" (Luke 24:35).

Emmaus is a place of epiphany, and on the hot, bright day when I walked into a monastery built on the site, I experienced that for myself. There were no tourists, but the church wasn't empty. Looking down from an upper loft, we could see nuns prostrate below us in total silence. I had never seen a nun before, but I was immediately overawed by what they were doing. What *were* they doing? I really didn't know, but as I watched them I fell as silent and still as they were.

Twenty years went by without remark on this memory, which is why I was even more delighted when my dad brought it up recently. "You remember that too?" I asked in surprise. The Wailing Wall, Holy Sepulchre, Jordan River, and Gethsemane – nothing made as vivid an impression on us as the nuns in Emmaus. And now I understand what they were doing that day: keeping silent adoration before the Blessed Sacrament where, like the disciples before them, they recognized and reverenced Christ "in the breaking of the bread."

A bench in the Old City of Jerusalem

These examples illustrate a point: children intuit meaning in things they can't yet fully understand. With the right teachers, they pick up more than good manners – they pick up good ideas. The nuns in Emmaus taught my dad and me something that all the buildings and sites we had visited had not. Their reverence was not a mental calculation; it was a full-bodied response to the truth they knew about who God is and how he dwells among us. I didn't have to understand all that to learn it.

We don't have to be young in years to learn reverence this way; we just have to turn and become as receptive as little children. For most of us, that will mean letting go of a certain cynicism, suspicion, and even dullness toward spiritual realities that we've learned from our highly developed consumer instincts: What am I being sold? How am I being manipulated? What can I get out of this?

> To learn reverence, we have to become as receptive as little children.

It's frightening to realize how deliberately irreverent our secular culture is. Alongside vast ignorance and inexperience there is something truly demonic at work here. Satan, after all, is the one who refused to serve. Irreverence is intrinsically disordered and, despite its seductive hold on our culture, basically ugly. If we cultivate it by choice, it chokes out the movements of charity, religion, and holy fear in us.

Understanding Reverence

Many of us, unfortunately, won't find good role models of reverence close by. But another way to learn reverence, which is open to all of us, is to grow in understanding.

I once met a man in his sixties who complained to me about the simplistic way his third-grade teacher had explained heaven to him. She told the class that they should imagine having good things in heaven just like they have good things now. For example, if they lived in a blue house, they would have a blue house in heaven. This analogy, similar to the one C. S. Lewis uses in *The Last Battle*, didn't sit well with him as a child: his family lived in an apartment!

"Now, isn't that a silly way to explain heaven?" he asked me.

"Yes," I agreed. "But you're not a child anymore. What have you done since third grade to understand heaven better?"

His is a common mentality. Many people grow up enough to criticize their first, childish ideas about faith, but don't follow through to more mature ideas about God, creation, and the human person. Left in a state of arrested development, reverence can't ripen.

Understanding allows us to experience reverence in its mature mode. It is the necessary completion of childlike imitation, no matter how young or old we are when we begin to comprehend divine truths. Understanding is something we can cultivate ourselves, firstly by studying the truths of nature and revelation, and secondly by removing misconceptions that are obstacles to reverence in our lives. If I imagine that God is a crotchety old man in the sky, for example, and don't know that he is spirit infinitely perfect and blessed in himself, who has created me to share in his goodness, I won't be able to assume an authentic posture of subjection to him, one of gratitude and supplication.

The point is that when I walk into a place of sacred beauty, the ideas I bring with me really matter. Irreverence grows out of ugly ideas I have about creation, the human person, and

God's saving providence. These misconceptions make it difficult for me to recognize beauty no matter how much of it I see around me. Becoming aware of my own misconceptions can open me to the experience of beauty in its three traditional properties – proportion, integrity, and clarity – and serve as a remedy for irreverence itself.

Reverence in a Flat World

One misconception that keeps us from experiencing the beauty of creation is radical egalitarianism. The culture around us endorses total equality in the worst sense of the word – we prefer to live in a flat world. The idea that I should submit to the excellence of another, God or not, is somewhat offensive. What we can't abide about the idea of reverence is its reliance on hierarchy, and we don't realize that an allergy to hierarchy is a reaction against a property of beauty itself, called *proportion*.

Proportion, as an aesthetic value, signifies the pleasure my mind takes in order and unity. We experience proportion in small ways every day: when we walk into a clean space instead of a cluttered one, or listen to a favorite song instead of the sound of traffic. On a much larger scale, however, the beauty of proportion is written into the universe itself.

That's because the universe has a form. Not a shape, mind you, but a principle that constitutes its proper perfection. It not only is something: it's also becoming something. And though we can't yet see entirely what God has in mind for it, the form of creation is already visible in the order and diversity of its parts. It shows us what it is by showing us the beauty of its proportions.

This instinct for the hierarchy in the universe is deeply Christian. Augustine, for example, paints the scene this way: "Where

Stairway in the Church of the Holy Sepulchre

does God's creation begin? With the angels. And where does God's creation leave off? With the mortal things of earth. The created order has an upward limit, beyond which is God; the created order has a downward limit, beyond which is nothing." Each being is placed. It is ordered. It is desired by God as one possible expression of his own infinite beauty among all others. But the point is that it belongs and it depends.

Depending on another is of the essence of hierarchy, and it's at the heart of created beauty. In fact, if you want a good example of dependent beauty, look at the human person in the attitude of reverence. We've already said that the first act of reverence is to submit the mind to God. But how?

The answer has to do with our place in the hierarchy of creation, and with the fact that we have both material needs and spiritual desires. It has to do with our capacity to grow in knowledge, love, and holiness. The answer is that reverence prays as Christ taught us.

Photograph by Mari. Used by permission.

Reverence and Bodies

Disregard for the body is a second common idea, one that keeps us from experiencing the beauty of the human person. Most of us are very comfortable with an easygoing attitude that says, "No worries, what I do with my body is spiritually irrelevant." The idea that bodies have responsibilities is unfamiliar, and the fact that reverence makes demands on our bodies seems rude. Reverence, religion, God: these are matters of the mind and soul.

Meanwhile, however, we're clearly missing something. And that's not a bad definition of ugly. Beauty is characterized by *integrity.* Integrity as an aesthetic value refers to the pleasure our minds take in seeing completeness and wholeness in things. It is the beauty of a thing that has all its parts, neither too many nor too few. It pleases me because through it I glimpse something of God – the fullness of being – himself. If I want to see the beauty of the world and my place in it (proportion), I can't overlook the fact that I am a soul *and* a body (integrity). It's precisely the integrity of the human person that explains why we owe God acts of embodied reverence.

We need our bodies in order to receive knowledge of the world around us. We derive knowledge from our five senses, all day long, every day of our lives. What's fascinating, however, is that we receive knowledge of God in the same way. God has chosen to reveal himself to us according to the embodied mode of the human nature he has given us – incrementally, over time, using sensible words and deeds that are often mediated through other people.

The bodily mediation of Christian revelation is a basic premise of the Incarnation: the Word became *flesh* to reveal the face of the Father whom no one has seen (John 1:18). What's more, this mediation gives us a new

Church of the Holy Sepulchre

Prayer is proper to someone in need, someone who depends. By prayer we accede to our place in creation and open ourselves to receive what God wants to give us as his children, *made* in his image and *remade* by baptism (Gen. 1:27; Rom. 6:3–4; Eph. 1:5). The hierarchy here is all in our favor, because lesser things are perfected by higher ones. What we gain by reverent prayer is nothing less than our own perfection. Listen to Saint Thomas Aquinas: "By the very fact that we revere and honor God, our mind is subjected to him; wherein its perfection consists, since a thing is perfected by being subject to its superior, for instance the body is perfected by being quickened by the soul, and the air by being enlightened by the sun."

In other words, just like life and light flow into the world from higher causes, so purity and firmness of mind flow into my soul from God when I revere him. This effect in me is called *sanctity*, and there is nothing, absolutely nothing, more attractive in creation. It is the beauty of being proportioned to divine things by grace and of sharing in Christ's own prayer to the Father who is "greater than all" (John 10:29). But in a flat world of low horizons, we miss this beauty.

perspective on the reverence we owe Christ in his humanity; it explains how what we pay him in justice returns to us in sweetness. Aquinas puts it this way:

> Matters concerning the Godhead are, in themselves, the strongest incentive to love and consequently to devotion, because God is supremely lovable. Yet such is the weakness of the human mind that it needs a guiding hand, not only to the knowledge, but also to the love of Divine things by means of certain sensible objects known to us. Chief among these is the humanity of Christ.

Christ's humanity is a divinely foreseen aid, a "guiding hand," as Aquinas says; our hearts and minds are weak and cannot pierce the heavenly realms on their own. This is why reverence for Christ's humanity – the kind of devotion motivating centuries of pilgrims to the Holy Sepulchre – is not an awkward compromise with superstition or an unfortunate falling away from pure, reasonable religion. Instead, it is an aid intrinsic to God's *ordinatio*, or saving order: "God," writes Saint Teresa of Avila, "desires that if we are going to please Him and receive His great favors, we must do so through the most sacred humanity of Christ, in whom He takes His delight."

That God continues to work within this bodily mode is a basic premise of Christian anthropology for which we should be grateful. It not only means we can hear God in Christ. It also means we can speak back.

Our bodies speak reverence to God in the same way they have been spoken to: incrementally, over time, using sensible words and deeds often mediated through other people. For example, when we humble our bodies in God's presence by bowing, kneeling, or even just folding our hands, we are using our bodies to speak reverence. When we deny our bodies by fasting, almsgiving, and other forms of Christian sacrifice, we are using them to speak reverence. When we gather our bodies at the times and places consecrated for God with the people of God, we are using them to speak reverence. And when we put our bodies at the service of others out of love for God and neighbor, we are using them to speak reverence.

At the same time, we are also shaping our lives into an integral whole that is nothing less than beautiful. But in a world where bodies have no responsibilities, we miss it.

Reverence and Providence

A third mindset does possibly more to mar our experience of beauty than the others because it is a misunderstanding about God himself. I'm referring to the habit of fatalism we've picked up from an unsavory friendship with materialism. When things – and their meaning – are reduced to matter and the laws of physics alone, what I see is what I get, and what I can't see can't mean much.

Fatalism like this blinds us to the actions of God in the world, and it makes the meaning of our own lives incomprehensible. Our

Real faith is not blindness; it is a supernatural way of seeing.

heads may be full of the latest news, but our minds can't discern the patterns of providence in these events. Meanwhile, the idea that we should revere a loving God who lets unloving things happen in the material world is hard to swallow. We doubt the beauty of divine providence because we don't see it in all its *clarity*.

Clarity is the third property of beauty – its illuminative quality – that manifests the inherent radiance of all beautiful things. As an aesthetic value, clarity represents the pleasure our minds take in finding light and intelligibility in the world. Clarity is at work, for example, when we see a stained glass window lit up and understand the story the window tells.

Clarity outstrips both proportion and integrity in the delight it causes, in much the same way that love outpaces faith and hope: It first recapitulates and then surpasses them. But clarity also excels in mystery. As beauty goes, it can be very *unclear* indeed, especially if I'm used to sights that are cheap and flashy. Real clarity is neither of those things. As Jacques Maritain explains, "To define the beautiful by the radiance of the form is in reality to define it by the radiance of a mystery. It is a . . . misconception to reduce clarity *in itself* to clarity *for us*."

The gap between what is clear *in and of itself* and what is clear *for us* is where the great mysteries of providence play out. It's also where our attitude of reverence must find its perfection in the supernatural virtue of faith if it is to survive at all. Reverence perfected by faith shows us the inner light of God's providence in everything that exists. With Saint Catherine of Siena, faith can say, "Hate sin and hold all else in reverence."

It's a common mistake to think that faith means closing our eyes and gritting our teeth against hard truths. Real faith is not blindness; it is a supernatural way of seeing that begins with the habit of believing God. Faith has God himself as its object, and it participates in God's own vision of the world. It enjoys a clarity that surpasses our natural powers.

Take, for example, the beauty of the cross of Christ. Materially, we can point to a brutal and humiliating form of execution that is very dark, indeed. With the eyes of faith, however, we can see the cross differently. We see the material form of an uncreated love, infinity in flesh, which is the light and salvation of the soul. We see a mystery of divine love, "a plan for the fullness of time, to gather up all things in [Christ]," a mystery which is the light and salvation of the world (Eph. 1:10). We see a mystery of human love, gratitude and surrender to the Father, which is the light and salvation of all the suffering in our lives. And seeing all this, we can exclaim with the Psalmist, "but there is forgiveness with you, so that you may be revered" (Ps. 130:4).

By faith, we behold already what heaven will show us in absolute clarity: that the mystery of the cross has been the singular beauty of our lives. But in a world darkened to divine providence, we miss it.

God Is Near Us

Reverence opens our eyes to beauty in the order of creation, in the wholeness of human persons, and in the mystery of divine providence – beauty that is inaccessible to irreverent eyes. As an attitude that accompanies Christian life, reverence is not optional but necessary.

Do Christians also need beautiful buildings, liturgies, art, and literature in order to convert the world? Absolutely. Even more importantly perhaps, we need them in order to convert ourselves. But there's an important sense in which each of us must cultivate a pilgrim's eye for beauty in the places in which we already live and worship and in the daily crosses we already carry. That means recognizing God's presence with us, in Christ, and submitting to him gladly through our practice of charity, religion, and holy fear. Once we begin to do that, our aesthetic experiences will show us more and more clearly that God is near us.

As a cloistered nun who no longer kneels in the great cathedrals, I can tell you that this kind of conversion by beauty is still possible, here in my own sacred space, with all its monastic simplicity. ➤

A Book to End All Walls

AN INTERVIEW WITH UK-BAE LEE

While adults debate border walls, one artist is planting seeds in the hearts of children who, he believes, will one day tear them down. This March *Plough* releases an English edition of Uk-Bae Lee's picture book, *When Spring Comes to the DMZ*. With the Korean Peninsula making headlines once again, we asked Lee for an update.

Artwork from Uk-Bae Lee's book

Why a children's book about the DMZ?

With other authors from Korea, China, and Japan, I started the "Picture Book Peace Project" to help children envision a more peaceful world. But when I heard the word "peace," into my mind flashed the brutal scenery of the DMZ, which is of course the opposite of peace, an image of war. The DMZ is like a wound scratched by a giant monster across the back of the Korean Peninsula, dividing South and North with sharp images of razor wire and ochre-colored earth that bring to mind the blood that has flowed there.

So how demilitarized is the DMZ really?

"Demilitarized zone" means an area free of weapons, but in actual fact there are still hundreds of thousands of landmines buried there and soldiers armed with high-tech weapons on either side. However – and this is a paradox – without the meddling of people the flora and fauna within the DMZ has flourished. The area has even been called a paradise for endangered plants and animals. It is, more accurately, a refuge of last resort. Still, we can see the irony that nature has benefitted from the troubles of humankind. The DMZ should make us think and reflect not only on the relationships between people, but also on the coexistence of humans and nature.

You visited the DMZ to create this book. What was most memorable?

The red-crowned crane family and the wild geese flying south over the barriers are unforgettable. The scenery was real, but I had a strange feeling of unreality: with such a tragic human history, how could the landscape be so beautiful?

What has changed in the DMZ as the result of recent meetings between the leaders of the two Koreas?

The gates are still closed and there is no freedom to travel back and forth. But some guard posts on both sides have been demolished and preparations are underway to reconnect the railroad and highway. The progress has been nothing short of a miracle, and people who love peace should continue to encourage these efforts so the seventy-year division of North and South Korea can end and an era of peace can begin.

At the end of the book, we see an old man and his grandson throw open the DMZ's gates and embrace their family from North Korea. When will that day come?

A peaceful world where separated families will meet and people can come and go freely will certainly come, because eighty million Korean people in North and South and overseas are earnestly wishing for peace. Perhaps the time of peace has started already, but a lot of work needs to be done to achieve "a peace that cannot be turned back from" and a world where war is no longer a threat. One picture book cannot change the world right away, but if it can leave a small impression, if it can move hearts just a little, then couldn't those hearts together change the world bit by bit?

In writing your book, were you also thinking of other walls such as the Gaza Strip barrier and the US–Mexico border?

Walls become firmer as more people despair of an end to the division. If we want to break down the walls, we need to first bring down the walls in our own hearts. A physical wall cannot be built in one morning or come down overnight, but the walls in the heart can come down all of a sudden.

Once we bring down the walls in our hearts we can start something small to help knock down the real walls. There is a Korean saying that "constant dripping wears away stone." The raindrops continually dripping down from the roof make a hole in the stone terrace. If we do not give up our dream for peace and are persistent in our efforts, the real walls will come down eventually. ⤳

Interview conducted and translated by Chungyon Won, January 3, 2019.

(continued from page 96)

He was thinking cappucinos and pastries. She had other ideas. Shortly afterward, Café Nicholson opened at 58th and Third. The menu was simple, delicious – and all Edna. Choice, she thought, was overrated. She would make one main dish a day, with sides and dessert. People would eat what was put in front of them. And it would be so good that everyone would enjoy it.

She also owned 50 percent of the place.

Café Nicholson, bohemian New York with a southern twang, was a huge success; regulars included William Faulkner, Marlon Brando, Eleanor Roosevelt, Truman Capote, Tennessee Williams, Greta Garbo, Salvador Dalí, and Gore Vidal.

Lewis's way of thinking and writing about food – she wrote the book on a series of yellow legal pads, which her niece typed up – was something entirely new. We would now call these recipes "farm-to-table." Each time of year had its proper food, and knowing how to cook was partly knowing when to cook. This kind of deep knowledge was what she wanted to preserve.

Lewis published four cookbooks in all, and went on to work as a chef in several prestigious restaurants, including Gage & Tollner in Brooklyn. She started the Society for the Revival and Preservation of Southern Food, and over the course of her career won numerous awards including, in 1995, the James Beard Living Legend Award. She died in 2006.

"As a child in Virginia," she told the *New York Times* in 1989, "I thought all food tasted

"It has been my lifelong effort to try and recapture those good flavors of the past."

Edna Lewis

delicious. . . . It has been my lifelong effort to try and recapture those good flavors of the past."

She did so – and in the process she recorded the community that those flavors were born in. "I grew up," she said, "among people who worked together, traded seed, borrowed setting hens if their own were late setting. . . . If someone borrowed one cup of sugar, they would return two. If someone fell ill, the neighbors would go in and milk the cows, feed the chickens, clean the house, cook the food, and come and sit with whoever was sick. I guess rural life conditioned people to cooperate with their neighbors."

Her attention to the subtle details of well-prepared food was only one aspect of her attention to the world from which that food came. "You felt all through her writing," said her editor, "that she was giving thanks for something precious." As she once told an interviewer, "I grew up noticing." In *The Taste of Country Cooking*, she describes a southern spring: "A stream, filled from the melted snows of winter, would flow quietly by us, gurgling softly and gently pulling the leaf of a fern that hung lazily from the side of its bank. After moments of complete exhilaration, we would return joyfully to the house for breakfast." ⟶

Sources. Edna Lewis, *The Taste of Country Cooking* (Knopf, 1976); Sara B. Franklin (ed.), *Edna Lewis: At the Table with an American Original* (UNC Press, 2018); Joe Rendall (interviewer), "An Interview with Chef Edna Lewis," video (YouTube, posted May 18, 2009); Bailey Barash (director), "Fried Chicken and Sweet Potato Pie," documentary film (*bbarash.com*).

Edna Lewis

JASON LANDSEL

"My first memory of who I was was food," Edna Lewis told a historian in 1984. She had learned cooking by watching her mother, her aunt, and the other women in their small community cook on wood stoves, she said. She learned to tell if a cake was done by listening to it – "the liquids make bubbling noises." She learned to use the cool well-houses and streams for refrigeration.

The inhabitants of the Virginia village of Freetown grew and harvested vegetables and fruit. They gathered berries, caught fish, and hunted game from the surrounding woods and fields in season. They grew and threshed their own wheat. Lewis's grandparents, along with others of the older generation, had received deeds to their plots of land after Emancipation in 1865.

The first thing that generation did, she said, was plant an orchard. They were planning for the long haul, planning to stay on that land. The meaning of the town's name was double: they were no longer slaves, and they were also free farmers, not tenants or sharecroppers.

"Their spirit of pride in community and cooperation in the work of farming is what made Freetown a wonderful place to grow up," she would write later in *The Taste of Country Cooking*, the cookbook that made her famous. "The farm was demanding but everyone shared in the work – tending the animals, gardening, harvesting, preserving the harvest, and, every day, preparing delicious foods that seemed to celebrate the good things of each season."

In 1976, when she was approached by Julia Child's editor to write the book – part memoir, part country-living manual, part cookbook – she had spent many years away from Freetown. After her parents died, she'd moved in the 1930s, as a teen, to New York City – part of the Great Migration.

> Lewis's way of thinking and writing about food was something entirely new.

"When I was a girl, they used to hang black men," she told an interviewer once, bluntly. "You couldn't do anything about it because they'd kill you. It scared the life out of us." In New York, Edna faced different kinds of prejudice: the limited work opportunities, the daily slights: having to use a separate entrance door to her job.

She joined the Communist Party, she said, "because they were the only ones who were encouraging the Blacks to be aggressive, and to participate. They gave me a job typing." It was there, too, that she met her husband, Steve Kingston, another Communist activist.

In the Party, she found a sense of the community that she'd left behind. She also found herself using the skills she had learned in Freetown at weekend CPUSA dinners. Johnny Nicholson, a fellow party member (or at least fellow-traveler) who came to those dinners, told her that he was planning to open an Italian-style café, and asked her to cook for him.

(continued on preceding page)

Jason Landsel is the artist for Plough's *"Forerunners" series, including the painting opposite.*